Henry Fairfield Osborn

Dedicated to my parents
Henry and Elizabeth
and to my sister
Celeste

Henry Fairfield Osborn

Race, and the Search for the Origins of Man

BRIAN REGAL

Ashgate

Published by

Ashgate Publishing Limited
Gower House
Croft Road
Aldershot
Hants GU11 3HR
England

Ashgate Publishing Company
131 Main Street
Burlington
Vermont, 05401–5600
USA

Ashgate website: http://www.ashgate.com

British Library Cataloguing in Publication Data

Regal, Brian
Henry Fairfield Osborn: Race and the Search for the Origins of Man.
1. Osborn, Henry Fairfield, 1857–1935. 2. Paleontologists—United States—Biography. 3. Paleontology. 4. Human beings—Origin. I. Title.
560.9'2

US Library of Congress Cataloging in Publication Data

Regal, Brian
Henry Fairfield Osborn: Race and the Search for the Origins of Man / Brian Regal.
p. cm.
Includes bibliographical references.
1. Human beings—Origin. 2. Human evolution—Philosophy. 3. Human Evolution—Religious aspects—Christianity. 4. Osborn, Henry Fairfield, 1857–1935. 5. Paleontologists—United States—Biography. 6. Eugenics—United States—History. 7. Physical anthropology—United States—History. 8. Racism in anthropology—United States—History. 9. Religious fundamentalism—United States—History. 10. Scopes, John Thomas—Trials, litigation, etc. I. Title.
GN281.R4123 2002
569.9–dc21 2001046039

ISBN 0 7546 0587 6

This book is printed on acid free paper.

Typeset by Manton Typesetters, Louth, Lincolnshire, UK.
Printed and bound in Great Britain by MPG Books Ltd, Bodmin, Cornwall.

Contents

List of Illustrations

Acknowledgements

This project began as a childhood fascination with dinosaurs, cavemen, exploration and esoteric ideas. I first seriously tackled Osborn and his human evolution theories as a beginning graduate student at Drew University in the history of science class of Darwin scholar David Kohn. As with any historical research project I have many people to thank, because without their help this study would not have been possible.

I would like to thank David Kohn, Donald Dayton and Charles Wetzel of Drew University, as well as Garland Allen of Washington University, St Louis, whose help and advice were invaluable. Others who read all or parts of the manuscript and gave useful suggestions include: Michael Ruse of Florida State University, Jonathan Rose and Brid Nicholson of Drew, Edward Rafferty of the Mary Baker Eddy Library, and Jonathan Spiro of the University of California, Berkeley. Crucial advice and counsel came from Mark Lender, Frank Esposito and James Jandrowitz of Kean University. Also, help from many archivists and librarians made this project much easier than it would otherwise have been. Pivotal assistance came from Barbara Mathe and her staff at the American Museum of Natural History's central archive. Also thanks to everyone at the New York Historical Society; the Pierpont Morgan Library; the archive of the New York Zoological Society; the Rare Book and Manuscript Department of the New York Public Library; the Manuscript Division, Department of Rare Book and Special Collections and Mudd Library at Princeton University (materials from their collections are published with the permission of the Princeton University Library); the New Jersey Historical Society; the Clements Library at the University of Michigan as well as the Pickler Memorial Library, Truman State University; the Bancroft Library, University of California, Berkeley, and Robert Cox at the American Philosophical Society of Philadelphia. Josie Cook and her staff at Drew University's Rose Library inter-library loan department and New York rare-book dealer David Bergman proved invaluable in obtaining hard-to-find articles and volumes. Also many thanks to John Smedley and everyone at Ashgate Press.

Also thanks to Tin Tin, the Tracey brothers and the late Jonny Quest. And of course Lisa Nocks, who suffered through it all, did important early editing, and helped make life worth living.

Notes on Sources

Abbreviations

APS	American Philosophical Society
CAAM	Central Archives, American Museum of Natural History
MAAM	Mammal Archives, American Museum of Natural History
NJHS	New Jersey Historical Society
NYPL	Rare Book and Manuscript Department, New York Public Library
OSNY	Osborn Papers, New York Historical Society
PU	Rare Book & Manuscript Department, Princeton University
MUDD	Princeton University Archives, Seeley G. Mudd Manuscript Library, Princeton University
TRUMAN	Harry Laughlin Papers, Pickler Memorial Library, Truman State University
VPAM	Vertebrate Paleontology Archives, American Museum of Natural History
VSOMorg	The Virginia Sturges Osborn Papers, Pierpont Morgan Library

Other manuscript material came from the Clements Library, University of Michigan, and the Bancroft Library, University of California, Berkeley.

Deconstructing Henry

In 1922, with the blistering 'Outer Mongolian' sun beating down on him, Roy Chapman Andrews stood on a bluff overlooking the vast Gobi Desert. In his fedora and leather jacket, a revolver hanging rakishly from his hip, Andrews could have been the model for the fictional Indiana Jones. Standing next to him in a stylish hunting jacket and pith helmet was Andrews's boss, Henry Fairfield Osborn, president of the American Museum of Natural History. Grouped around them their support team – over two dozen American, Chinese and Mongolian geologists, paleontologists, translators, engineers and camp technicians – stood like a dust-covered Greek chorus. Behind them stretched a fantastic lifeline of cars, trucks and a hundred-camel caravan. All this was supported from Peking, China by a line of local bureaucrats, officials, princes, warlords, and wealthy American philanthropists and industrialists that ran back to New York. They had come halfway round the world on the most highly publicized scientific expedition in American history. Their arrival here was the culmination of twenty years of scientific inquiry on Osborn's part. He hoped it could be the fulfillment of what for him was the most important question in human existence: where did we come from?

The valley in front of them was covered in fossil bones. Millions of years of erosion had uncovered a treasure trove of extinct animal remains. The material they were sending back to their museum was revolutionizing the scientific community's understanding of the ancient world, making 'dinosaur' a household word, and initiating a craze for the creatures that has continued to the present day. Headlines around the world heralded the expedition and millions followed every installment of this great romantic adventure.

Though today it is synonymous with the discovery of a fantastic array of dinosaurs – including the first-known dinosaur eggs – the Central Asiatic Expedition was in large part an attempt to prove that Central Asia was the cradle of man. Andrews had become fascinated with the theory of human origins developed by Osborn, and he was determined to help prove it. Andrews's mentor believed human evolution was purposeful, had specific direction and was built upon divinely inspired laws of organic change. He argued that the present human races first appeared from human-like ancestors with little direct connection to the primates. The difference was in their level of struggle. The primates lived in environments that were luxurious and placid, while the humans lived in a harsher, less forgiving climate. Osborn said that man's 'struggle for existence was severe and evoked all the inventive and resourceful faculties he possessed.'[1] These early 'Dawn Men' began to evolve into more advanced creatures because they used the inherently superior characteristics of their 'race plasm' (a pool of hereditary material) to overcome the harsh environment they were born into. The population that had the greatest race plasm – and therefore advanced the furthest – became the Aryans, Nordics and Anglo-Saxons. It was the beginning of the great human adventure that was not just the struggle between man and environment, but between predestination and free will. 'Nothing in this world', Osborn said, 'would be gained without effort.'[2] He claimed the ultimate effort, as well as the 'drama of the prehistory of man,' had begun here in Central Asia millions of years before.[3]

Osborn considered himself a staunch empiricist whose methodology was to discover the truth about the past by studying the fossil record. He saw the evolution of life in general, and human evolution in particular, as history. The fossil record constituted the bits and pieces of the past which, collected and collated in proper Von Rankean fashion, would fall into place of their own accord and give the clear, objective and unambiguous story of the past, along with its inherent truths, morals and certainties (though he then wrote that history down as an almost romantic form of literature with just the right happy ending). Osborn had been constructing his theory since the turn of the century, but he was unable to search Asia for the supporting evidence until Andrews's adventurous zeal finally made that possible in 1920. He was here to search for the first humans, but in a way, Osborn was looking for himself. The two men glanced at each other, smiled, then hurried down the bluff like boys let loose in a toy store.

As president of the American Museum of Natural History, Henry Fairfield Osborn (1857–1935) was an important figure in late nineteenth- and early twentieth-century American science. His career put him in a unique position to pursue his quest to find the first humans. Because of his public profile, Osborn's pronouncements on science, education and the state of society were taken as authoritative by many. In his day he was second only to Albert Einstein as the most popular and well-known scientist in America. Osborn's position, his extensive publishing in books, periodicals, journals and newspapers, his role in immigration legislation and the eugenics movement, science education, and controversies surrounding questions of human evolution and its impact upon America, make it essential that the intellectual basis of his work be examined.

Though the racial elements of American science, particularly the eugenics movement and racial anthropology, have been studied extensively – in the work of George W. Stocking, Daniel Kevles, Diane Paul and others – generally the topic of race and human origins is dealt with as part of a wider discussion of racial attitudes in evolution, such as in John S. Haller Jr's *Outcasts From Evolution* (1971), Nancy Stepan's *The Idea of Race in Science*

Fig. 1 'The Professor' Henry Fairfield Osborn at the height of his powers as president of the American Museum of Natural History in New York.

Neg. No. 109272, courtesy Department of Library Services, American Museum of Natural History.

(1982), and Edward Larson's *Sex, Race, and Science* (1995), or only in the modern context, as in Wolpoff and Caspari's *Race and Human Evolution* (1997). None of these works addresses the role of Henry Fairfield Osborn and human origins. Peter Bowler has done extensive work on the history of evolution theory, but because of its wide scope, his *Theories of Human Evolution* (1986) gives Osborn only cursory mention. Historians tend to concentrate on issues of human evolution that involve questions of natural selection, species formation, whether evolution was thought to be progressive, or more often in the context of social origins, the origins of human intellect, or early human culture. However, in the history of evolution, the question of where the first humans appeared played a subtle but important role in the overall question of race and human evolution – particularly when it was injected into political and social concerns.

Race as a biological idea can be traced to the beginnings of anthropology in the 1700s. As Europeans encountered different people around the world their curiosity about them increased. Biological systematizers like Carolus Linnaeus and Johann Friedrich Blumenbach sought ways to work these people into an orderly pattern which itself could be placed into the long list of organisms that inhabited the earth. This attempt to arrange organisms allowed for hierarchies based on various rules of complexity and appearance. While early attempts to arrange human beings were innocent, eventually, as race became a biological question, it took on a darker aspect. Close studies of human anatomy and behavior led to the quantification of human characteristics so that by the late eighteenth and early nineteenth centuries the classification of people to show their similarities had evolved into a method to separate people by their differences. With the increasing sophistication of biological science and the appearance of Darwinian evolution theory, perceived racial differences were used as indicators of their relative superiority and inferiority. With the growing acceptance of science as an intellectual authority, racial distinctions became biological 'facts' which could be used to support and justify social and cultural behavior from simple prejudice to slavery, imperialism and mercantilism. It was largely a result of the work of Henry Fairfield Osborn that the origin of humans was linked to questions of evolution, the biological study of race and ethnicity, and social Darwinism in a way that it had not been in the past. Part of Osborn's larger objective was to show that an individual's characteristics were a result of biological and geographical origins. He believed that by explaining the origins of different groups, society would have a guide as to how to treat those groups. It was already common to treat people in accordance with their ethnic origins. Osborn's search for the origins of man was this idea taken to its biological extreme.

There have been few examinations of Osborn's life and work. William K. Gregory's 'Biographical Memoir' (1938) is more hagiographic than critical, William Berryman Scott's is the recollections of a best friend, and Edwin Colbert's several entries on Osborn in various publications have been all too brief. The only full-length historical work on Osborn is Ronald Rainger's excellent *An Agenda for Antiquity* (1991). Rainger's book is less a biography than an institutional history of Osborn's key role in making paleontology a central discipline in the study of evolution, and in turning the American Museum of Natural History from a dusty Victorian cabinet of curiosities into a leading center of scientific research. While he does discuss Osborn's views on race and society, for Rainger, the 'Agenda' of the title was Osborn's commitment to use science as a way to promote the ideals and politics of the power élite of New York. By promoting science and the study of nature as a recuperative elixir for the breakdown of American society, Rainger argues, Osborn could help bring about respect for law and order amongst the lower classes, reverse the ravages of the Industrial Revolution and the modern age, and preserve Nordic–American control of the corridors of power in the face of massive immigration and social unrest. While Rainger establishes this central point, he does not address the deeper personal motivation that informed Osborn's work. Indeed, that personal drives and demons were, in some cases, a catalyst to human evolution studies is an under-examined aspect not only of Osborn's work, but also in the historiography of evolution.

For Osborn, politics was important, but there was a deeper element which engaged his thinking. As a youth he was taught by his devout Christian mother that to struggle beyond what one was born with – wealth, intelligence, social position and such – was key to achieving salvation. Virginia Osborn believed that it was not enough to rest on advantages provided by birth. For her, advance could only be gained by going beyond those advantages to reach a higher spiritual level. I argue that, as a scientist, Henry Fairfield Osborn subconsciously used his mother's salvation credo as part of the foundation of a theory of evolution. However, his work was not cut out of her motherly sayings and advice alone. Working with James McCosh and Arnold Guyot of Princeton, and particularly Edward Drinker Cope of Philadelphia, Osborn began to meld religious conviction and scientific theory as they did. This secularizing process led him to be convinced of the validity of his position. He argued, in a nuanced combination of science and metaphysics, that evolutionary advance was only possible if organisms struggled against the environment to go beyond their inborn biological characteristics. This slowly manifested itself in what became, by 1900, the central program of his career: linking human origins to racial hierarchies to support his conclusions on evolution. If he could demonstrate through the fossil record that humans had risen and evolved because they struggled beyond their hereditary characteristics, that evidence would in turn support his metaphysical credo.

Osborn's belief in the Central Asian origins of humans, his role as an activist for eugenic reforms and immigration restriction, his support for Nordicism, questions of social order and control, his place in the 'new' versus 'old' biology debate over the changing approaches to the study of living things, his acceptance or rejection of various human fossils, his role in the Christian fundamentalist controversy with its science versus religion battles and Biblical literalist rejection of Darwin, the Scopes 'monkey trial' concerning the attempt by Christian Fundamentalists to curtail the teaching of evolution in schools, and finally his construction of the 'Dawn Man' hypothesis, all of the ideas he developed, the theories he put forward and the pronouncements he made were linked to his desire to confirm and support his evolution theory. But Osborn did not think that simply proving his theory of evolution would save his soul. It is my contention that Osborn believed that if his theory of evolution were borne out it would, by analogy, suggest that his belief in the possibility of achieving salvation would be borne out, since both were products of God's laws.

It is accepted wisdom that Osborn's thinking on science and evolution was colored by his racism (he certainly held opinions that today would be considered racist and anti-Semitic). This view is a bit simplistic. As with his political ideas, his views on race were a result of a complex interaction between his science and his religion. A different interpretation can be considered if we look to these deeper motivations. His study of the fossil record had suggested to Osborn that as mammals became more specialized in their adaptation to specific environments they lost their ability to overcome environmental change: they lost the ability to struggle for advancement. It was this loss of ability that, to Osborn, made highly specialized forms inferior. A more generalized species could adapt to a wider range of circumstances and advance further along the evolutionary scale. This made more generalized species superior. Having first worked out this notion for mammals, he applied it to humans. He saw Nordics as a more generalized species than Africans or Asians.

Raised as a Presbyterian, with a healthy dose of Scottish Common Sense philosophy, Osborn held strongly to a basic set of Christian values and morals. He regularly attended church throughout his life (both alone and with his family), but was not a doctrinaire. He held the Bible as the basic guide for living a Christian life, but balked at slavishly following any particular denomination's edicts or rules. With the exception of the fundamentalist controversy over evolution, he never engaged in religious politics, preferring to keep his deepest spiritual views private and never articulated his own notions of the meaning of salvation. Although he had many friends and associates who were theologians, he rarely discussed theology openly. Considering himself outside such squabbles, he claimed that, 'I am not tongue-tied by adherence to any denomination, creed, or dogma.'[4] He fought his battles over predestination and the efficacy of good works in the realm of biology.

In his *Social Darwinism in American Thought* (1955), Richard Hofstader argues that Liberal Protestants of the later nineteenth century used science and evolution to construct an alternative naturalistic Calvinism. In this view man's adversarial relation to God was replaced by a similar relationship to nature.[5] Building on this assertion, James Moore, in his work on the reception of the teachings of Herbert Spencer in America, adds that this naturalistic

Calvinism 'furnished a systematic and transcendental justification ... for law and order, and hope.'[6] Osborn followed a similar route. Although he was a deeply religious man, he transferred many of his religious beliefs and yearnings on to science, and sought answers from science normally asked of theology.

As part of his work Osborn developed a resistance to the idea that humans and primates shared their origins or a close biological relationship. He openly banished the primates and fossil humans from direct human ancestry in the mid-1920s. To him they seemed degenerate evolutionary dead ends, not fit to be part of the progressive human line or its optimistic history. In the fundamentalist controversy, Osborn fought to preserve evolution by arguing against the close relationship of humans and primates. Ironically, he helped the fundamentalist cause (which he opposed) by inadvertently seeming to give the stamp of approval to their anti-evolution, anti-ape position.

Osborn constructed his theory of human evolution at a time of considerable intellectual debate over the meaning of evolution and its impact on society. As many scholars have shown, while the United States in the late nineteenth and early twentieth centuries was predominantly Protestant Christian, and contained several large powerful denominations, it was far from a unified front. Protestant orthodoxy was under pressure from many quarters. Evolution theory and recent discoveries in geology took their place alongside biblical criticism, the rise of new sects and denominations (such as Mormons, Adventists and Christian Scientists), the influx of large numbers of Catholic and Jewish immigrants, the growth of liberal Christianity, as well as a general disillusionment with traditional church orthodoxy increasingly felt by average citizens. Doctrinal fights over issues such as the meaning of salvation, the Virgin birth, denominational autonomy and other such squabbles rent American Protestantism from North to South. Evolution only exacerbated an already tense situation.

Adding to this fragmentation is the problem of labels. Osborn's place in the fundamentalist controversy is made murky by the fact that such terms as 'fundamentalist,' 'modernist,' 'orthodoxy' and 'liberal' are at best generalizations. As Paul Conkin points out in his work on the Scopes trial, 'even to utilize such labels is already to engage major definitional issues.'[7] Indeed, labels in both science and religion from this period have problems of definition. For example, not everyone who considered themselves a Darwinist actually was one, and not everyone called a fundamentalist was one. Also, during the debate between the 'old' and 'new' biology, many involved did not fall cleanly into one camp or another. Osborn fought in a world where most foes were ill defined. Compounding the problem, few protagonists on either side ever realized it.

Osborn discovered early on that the attempt to support his evolution work would not be easy. What complicated the matter was that he was torn between the older world of Victorian naturalists and the new experimentalists of the twentieth century. The old ways comforted him; the new ones frightened him. Unfortunately, with the birth of the 'new' biology Osborn could not simply dismiss the experimentalists as some older morphologists did. He was scientist enough to know that the new men were on the right track, but was too chained to the baggage of the past to follow easily. The razor's edge he walked between the past and the future caused him great personal turmoil. His life is an example of how personal belief can affect scientific research. It is the drama of a man struggling to reconcile science and religion, of a man who realized that his most dearly held beliefs and certainties were slipping away, and who desperately tried to save them. He fought battles in academia and in public, but was often arguing with himself. His inner conflict was such that he can be seen as a tragic character left behind by the tide of time, never to achieve what he sought. He weathered the upheavals and clashes in science, religion, American culture and his own life, never losing sight of his goal. While the outcome seemed clear to him early on, he persisted in his efforts to work it out, never giving up or giving in. With single-minded purpose he constantly battled against every new obstacle with renewed vigor.

Osborn was not alone in his attempt to link the origins of the human race with questions of salvation, secularization, race and ethnicity. Others, from the *philosophes* of the Enlightenment to the mystics of the occult revival, to the proto-facists of the early twentieth century, were all looking for answers to similar questions. Significantly, they all looked to the same place for answers – Central Asia. The foreboding wastes of the Gobi Desert became the jewel that many human origin theorists strove to possess. It was also the starting-point for Osborn's version of the human adventure. Proving that the first humans appeared there would be the building block upon which he could erect his entire human evolution theory. That is why he so doggedly held to it, even in the face of accepted scientific wisdom and evidence.

Osborn has been characterized as a pompous, self-important character from a politically incorrect past. While that is a generally accurate appraisal, he is far more complex than this one-dimensional portrayal. He was a man of contradictions: a man who loved Charles Darwin, but was not a Darwinist; a deeply religious Christian who vigorously opposed Christian fundamentalism; a Nordicist and eugenicist who argued that all races had their geniuses who ought to be cultivated and promoted; a man who railed against the influx of immigrant children yet fought to improve the public school system to educate those very children; a man who believed that predestination ruled evolution and that personal motivation and will could alter the course of evolution; a snobbish patrician who promoted the careers of many of his underlings. These contradictions are the reason for the methodology employed here.

The aim of this study is to explore the manner in which Henry Fairfield Osborn constructed his human evolution theory; how he defended it, modified it, and tried to hold on to it in the face of scientific resistance and neglect. I will use the method of intellectual biography. All ideas are shaped and tempered by the life experiences of those having them. As a scientist, Osborn built his work on meticulously gathered evidence, but that evidence was analyzed and interpreted through the filter of his life and written up as a form of historical literature. The work of Misia Landau has shown how this literary approach was not uncommon for paleontologists at the time.[8] Intellectual history, as historian of anthropology George Stocking interprets it, 'is the history of men thinking rather than the history of thought.'[9] While it can never really be known what Osborn was thinking, interweaving a constructed narrative of his scientific development with the story of his personal trials and tribulations may help open up the progression of his thought process. And since Osborn constructed his ideas in a historico-literary way, it seems fair to approach the telling of the story of its construction in a similar fashion. Philosopher Michael Ruse (1997) has suggested that a common flaw in dealing with the history of science has been that works are often either externalist or internalist. He suggests that a combination of the two would yield the best results. This work will be an attempt to balance the internal with the external as a way of getting at the complex interplay of scientific thought and social experience, and the resultant ideas and theories.

One word about political correctness. I often use the expression 'man' in this study. It is considered proper today to use expressions like humankind, humanity and such. I use the older term because the subjects of this book used it.

Notes

1. Henry Fairfield Osborn, 'Why Central Asia?', *Nature*, 26:3 (1926), 263–269. From here on Osborn's name will be abbreviated as HFO in footnotes.
2. HFO, *The Earth Speaks to Bryan* (New York: Charles Scribner's Sons, 1925), 25.
3. Quoted in Roger Lewin, *Bones of Contention* (New York: Touchstone, 1987): 34.
4. HFO, *Science and Religion: Polemics in the Fundamentalist Controversy* (New York: Charles Scribner's Sons, 1926), 181.
5. Richard Hofstader, *Social Darwinism in American Thought* (Boston: Beacon Press, 1955): 10.
6. James Moore, 'Herbert Spencer's Henchmen: The Evolution of Protestant Liberals in late Nineteenth-century America', in John Durant, ed., *Darwinism and Divinity* (New York: Basil Blackwell, 1985), 76–99.
7. Paul Conkin, *When All the Gods Trembled: Darwinism, Scopes, and American Intellectuals* (New York: Rowman and Littlefield Publishing Co., 1998), 50.
8. Misia Landau, *Narratives of Human Evolution* (New Haven: Yale University Press, 1991).
9. George W. Stocking, ed., *Race, Culture and Evolution* (Chicago: University of Chicago Press, 1982), 5.

CHAPTER ONE

'The Period of Rash Guesses'

Henry Fairfield Osborn's Central Asia hypothesis, and his search for the origins of man, must be understood as part of a fascination with the idea of identifying the original home of the human race and using that knowledge to promote social, nationalistic and personal salvation agendas which flourished from the seventeenth through the nineteenth centuries. Human-origin theories could be used to further views on race relations and the status of non-white European peoples, as well as justification for nationalism, colonialism and European economic and cultural domination. It also served 'pseudo-scientific' ideas like eugenics, polygenesis, racial anthropology and even occultism. While Osborn shared affinities with many of the movements and individuals that dealt with human-origin questions, he is not their direct intellectual descendant. He was not attempting to fulfill their programs or carry on their legacy. His work sometimes runs parallel to theirs because his own research led him to similar places and sometimes similar conclusions.

Before addressing Osborn's search for the origins of man in particular, it is important to look at these earlier theories in general because they laid the groundwork over which he walked. Osborn was not the first, but the last, major thinker to look at Central Asia as the cradle of man. While it was not his intention to do so, in trying to prove that Central Asia was the home of the human race, Osborn finally helped show that it was not.

From the Enlightenment on, the secrets of man's first appearance on earth were pursued by a wide range of intellectuals in Europe, Great Britain and the United States, from anti-clerics to pro-slavery advocates, modernists, anti-modernists, naturalists, nationalists, romantics and occultists. Although they had diverse backgrounds and differed in their approaches, they shared a basic desire to use such knowledge to raise themselves above the ordinary run of men, to form an alternative to mainstream religion, and somehow confirm the promise of salvation. The search for human origins thus became less an attempt to discover the facts of humankind's past than a way of bolstering cultural and patriotic notions about the present, and of securing good lodgings in the afterlife.

In Osborn's hands the hypothesis would finally meet the test of empirical reality. Yet his own motivations for mounting and supporting the Central Asiatic Expeditions owe as much to the extra-scientific baggage of the Central Asia hypothesis as to Osborn's scientific commitments. Finally, it must also be remembered that not all human-origin theories were based on philosophy, metaphysics, or political agendas. Much solid scientific work went into the study. Some of the great scientists of the era turned their intellect to the problem of human origins. Their work was often used to bolster the assertions of the philosophers and mystics.

Underlying the complex motivations of human origin theorists were two practical problems. The first was a desire to remove religious dogmatism from the discussion of the natural world. The growing dissatisfaction with traditional religion and church teaching led to efforts to create an epic story of biblical proportions unconnected to Christianity, but promising the same outcome of redemption and triumph as well as social uplift.[1] The study of human origins was only a part of a wider trend in the secularization of science. Many areas in the study of natural history were moving away from religious justifications and concordances. Geology, physics and astronomy were all slowly but surely moving out from under the shadow of theological control.[2]

The other problem was to find a geographically suitable location for the birthplace of humanity. To the Western world the archetype of human origins was the story of the Garden of Eden. What was being reached for now was a new archetype. Attention began to move away from the Holy Lands of the Middle East to India, then to Central Asia. This shift coincided with the growing Western fascination with Asia, a place little known to Westerners. Its mystery provided a *tabula rasa* on to which they could write any history they wanted. As the birthplace of the human race, Central Asia became a substitute for the Garden of Eden because it seemed to solve many of the problems associated with human origins without demanding too much in return in the form of verifiable facts and realities. Asia and its culture offered complex and sometimes contradictory solutions to the intellectual and spiritual problems some felt were inadequately served by Western ideologies.[3] While a detailed study of the idea of Asia and 'The East' is beyond the scope of this work, it is important to understand that Asia became not only the cradle of man, but for some a source of substitute spirituality, though with a rational underpinning. 'The East,' as a theoretical construct, contributed to the foundation of a form of what M.H. Abrams calls 'Natural Supernaturalism': a spirituality based in reason and the laws of nature.[4]

This chapter will be a survey of the diverse story of movements, themes and individuals involved in the early search for the birthplace of humanity. It will look at the French Enlightenment, the Romantics, the German Neo-Romantics, British empiricists, the occult revival, and the pre-Darwinian notions of race and human biology in the United States. As a meta-idea the Central Asia hypothesis was sometimes a major theme and sometimes only a thread, but always a persistent one. This will help contextualize Henry Fairfield Osborn's pursuit of the hypothesis and his work on human evolution.

The Enlightenment and the Romantics

The modern search for the origins of humankind began in seventeenth-century Europe as a result of a search for national identity as well as a growing dissatisfaction with a literal reading of the Genesis story. Ethnology was as much political as it was scientific.[5] Establishing historically deep roots could anchor a people and give them continuity with the past and a sense of connectedness. By the late 1600s some French intellectuals claimed their people to be descendants of the Trojans. In 1703 the monk Dom Perzom wrote that the biblical Gomer was an ancestor to the Gauls as well as one of the Titans of Greek mythology, and that the home of the Titans was upper Asia. French geographers thought that high mountains would be the ideal locality for the first humans and so suggested the Himalayas of Tibet. Voltaire felt that Adam must have taken his culture and society from India; Diderot's encyclopedia likewise argued that the oldest science came from the sub-continent. Pierre de Sonnerat said in his *Voyages Aux Indes Orientales* (1782) that the birthplace of man was India, while Immanuel Kant argued for Tibet.[6] In his *Histoire Naturelle* (1749–67), the French natural historian Count de Buffon (1707–88) argued that man's birthplace must have been in a temperate zone because a good climate would breed good men. Using an aesthetic criterion when looking for early man – the more beautiful the man, the earlier his appearance – Buffon theorized that the region around the Caspian Sea and the Caucasus mountains would be the perfect place to look.[7] Anti-clerical feeling was high amongst many French philosophers and they searched for ways to take intellectual discourse out of the hands of the Church. With Asia the *philosophes* could stick their intellectual finger in the eye of clerics and thus delightfully annoy the church hierarchy.

Not everyone in France was ready to abandon the biblical basis of man's history, however. For example, Jean-Sylvain Bailly (1736–93) argued that the learning of classical Egypt and India was all that was left of an earlier and more advanced civilization that had its origins at the North Pole. In 1779 he linked the North Pole, Central Asia and the Bible by claiming that following Noah's flood, an advanced civilization moved out of the temperate paradise of the Arctic to the plains of Central Asia, and then spread out around the world.[8]

German naturalist Johann Friedrich Blumenbach (1752–1840) also believed in the Asian origins of man. Like Buffon, Blumenbach said the first inhabitants of the Caspian Sea region were light-skinned, and that racial variation was a result of the original light pigment turning dark. In addition to coining the term Caucasian, Blumenbach was the first to apply the word 'race' to humans. In 1775, he classified the five races of man as Caucasoid, Negroid, Mongoloid, Polynesian and American. He also made these distinctions based on aesthetic appearance and the geographic distributions of the people involved, and was inspired by the work of the Swedish systematist Carolus Linnaeus and his *Systema Naturae* (1758).[9]

As the Enlightenment gave way to the era of the Romantics, ideas about human origins changed. The Romantics, particularly in Germany, had become disillusioned with what they saw as the cold and mechanistic views of the Enlightenment and the growing industrial age, and wanted something deeper. They sought to bring spirit, emotion and passion back into philosophy, and saw the study of nature as the place to learn the secrets of the universe. Friedrich Schlegel (1772–1829) captured the imagination of a generation of German youth with his stories of the Germanic past. He continued the growing Indian mystique by asserting in 1805 that India was the point of origin for European culture. Schlegel, in helping to begin the German Romantic movement, linked language to race in a hermeneutic technique that would prove popular to later anthropologists. Schlegel learned to read Sanskrit and became a devotee of Asian culture. This led him to develop the notion of an Indian/Germanic race for which he coined the term Aryan in 1819.[10]

Two years later Johann Gottfried Rhode proposed Central Asia as the home of this Indo-Germanic people, while G.W. Hegel argued that Europe had been colonized by this superior race. Until the eighteenth century it was widely held that because of stories in the Bible, the cradle of man was the Middle East, and the first language Hebrew. 'The Bible functioned as a kind of Kuhnian paradigm for research in the cultural, linguistic and physical diversity of mankind,' but was quickly losing that position.[11] Some Europeans were growing uncomfortable with the Semitic origins of man; Indian origins were readily accepted because they seemed to link the Germans to a 'higher' culture, and a non-Jewish race.[12]

With little else to go on, human-origin theorists turned to language as a key to unlocking the mysteries of humanity's past. The emerging discipline of philology – the study of the origins of language – became an important technique for studying human origins. The earliest languages were linked to the earliest peoples; the study of one was thought to reveal secrets about the other. Language was seen by many philologists as key to understanding both anatomical and philosophical human development. The idea of an Indo-European language group developed in 1788 after Western linguists began to learn to read the ancient Indian writing system, Sanskrit. They thought it was closely related to Greek and Latin, and by the early part of the nineteenth century German grammarian Franz Bopp, in his *On the Conjunctive System of Sanskrit* (1816), was arguing that the first languages had appeared, along with the first men, in Kashmir (which he believed was the Garden of Eden).

Linguistics was important to the French author Ernest Renan as well. Best known for his *Life of Jesus* (1863), Renan speculated on human origins based on the relationships of various languages. He wanted to substitute philology and science for divine intervention, and saw language as the engine of human experience rather than God. Dissatisfied with conventional Christianity, Renan looked for a substitute in Asia. Renan's India, like that of a growing number of Asia theorists, was not contemporary India, but a mythical Aryan India. A racist anti-Semite, Renan separated people into superior and inferior groups and saw the past as an indicator of current possibilities. He argued that the structure of languages revealed the Aryan origins of mankind and that they were a superior race.[13]

The German philosopher Johann Gottfried Herder (1744–1803) opposed Noachian or Jewish ancestry for Europeans, especially Germans. He said, 'Asia prepared the abode of the first human race.'[14] Their admiration for Eastern culture and civilization prompted many German philosophers – like Herder, Goethe and Schopenhauer – to look there for the source of mankind. The German Romantics and Idealists, disillusioned with the seemingly soulless mechanics of the Enlightenment, looked to 'spirit' and 'will' as the driving forces of the universe. They ascribed strong will and spirit to the hypothetical Aryans and characterized them as physically, mentally and culturally superior. Idealists like Hegel saw history developing in an inevitable sequence of distinct stages driven by a spiritual force beyond the control of men, so he modeled the Aryan myth to fit that idea.[15]

By the middle of the nineteenth century, the assumption that Asia was the cradle of man, and the belief in the existence and superiority of the Aryan race were well entrenched in European thinking. The Frenchman Arthur de Gobineau's *Inequality of the Races* (1853) and *Moral and Intellectual Diversity of the Races* (1856) not only hypothesized that the Aryans were the world's superior race, but also that they originated in Central Asia. He was the first to articulate the idea that history was a racial struggle, and that to study it was a form of biology.[16] Gobineau believed that the lesser breeds of men appeared in the Middle East as the product of a separate creation. He feared miscegenation and opposed democracy because it allowed the inferior rabble to control the superior racial aristocracy. He also discounted (as many writers did) the effects of environment, stating that racial superiority and its characteristics were biological and hereditary.[17]

The Races

Beginning in the early nineteenth century, the central debate over human origins was not so much where, but how the different races had appeared. The debate broke down into two warring camps. The first – the unity of creation, or monogenist, school – argued that all humans were descended from the same stock. They believed that all men were at least biologically equal, if not always culturally or socially so. In opposition to this notion was the polygenesis school, which believed that each race had somehow appeared separately as the result of multiple creations. Whereas the monogenists thought all men nominally equal, the polygenists argued that the races were inherently different and separate, and thus not all entitled to the same treatment.

An important proponent of polygenesis in Germany was Ernst Haeckel (1834–1919). After reading the *Origin of Species* in German translation, Haeckel became a major advocate of evolution. The living embodiment of the Nordic ideal, Haeckel loved sports, nature and the Fatherland. Though a Romantic, he studied medicine and marine biology. His work on the microscopic life form *radiolaria* showed him that living things did go through a series of transitional stages, and so he embraced Darwin's ideas about transmutation and variability. The creator of the first genealogical tree of life, Haeckel coined the term phylogeny in 1862, and argued that the embryonic forms of a species showed the stages that an organism went through during evolution.[18] He also developed the idea of monism, which stressed man's mystical oneness with the universe. He warmed to evolution in part because of its political implications. He felt change was not only possible, but necessary for living things and history. An anti-cleric, Haeckel was an outspoken advocate of Aryan superiority and the mystical German past which he believed had been ruined by Christianity.[19] For Haeckel, not only did people have souls, but so did nations and races. He wrote popular books on the soul and human evolution, and in 1906 formed the Monist League to promote his ideas about German society, social Darwinism and eugenics, and his view that race and nationality were closely linked.[20]

Revolution and political upheaval reduced the German states to a condition of cultural fragmentation and confusion as they strove to unify in the nineteenth century. It was this fragmentation, however, that in part created the need for some unifying cultural or even biological history to pull the German people together. Haeckel embraced the Asian origins of man partly because it seemed to be, as it did for the earlier French *philosophes*, an ideal weapon with which to attack the Christian world-view. He combined science, politics and religion, and equated the steady march towards perfection he saw in nature with man's rise to perfection. Haeckel felt that knowledge of the history of evolution and human origins could bring man into a closer relationship with nature: one devoid of any clerical intervention. The British naturalist T.H. Huxley was so concerned with Haeckel's mix of science and extra-science that he cautioned the German against going too far with his speculations.[21]

Not to be deterred, Haeckel gave free rein to his mystical, occult proclivities and conjectured that the ancestors of the Aryans had come from the lost continent of Lemuria in the Indian Ocean.[22] Ape-like proto-humans, he argued, entered India from Lemuria and then proceeded to spread into Europe, evolving as they went. In *The History of Creation* (1879) Haeckel included a 'hypothetical sketch' of the migratory routes of the Lemurians. As they spread out around the world they evolved into their current racial forms, of which the Aryans were the superior.[23] Haeckel's science gave race theorists an authority to fall back on and apparent absolutes to save them from the ravages and degradation of the modern world.[24]

The first outspoken proponent of the unity school was James Cowles Prichard (1786–1848). Prichard was the first English investigator to study race scientifically and systematically. He dropped all references to the Bible and other vague speculations and based his work upon the anatomical studies of Blumenbach.[25] A competing bastion of polygenic belief was the Anthropological Society of London. Formed in 1863, the society was the brainchild of Dr James Hunt, who linked race, science and politics into a heady brew designed to justify racial discrimination against non-Anglo-Saxons. The American Civil War had Hunt agitated over the emancipation of black slaves and he decided to take action.[26] Hunt was a student of the British anatomist Robert Knox, who held that race was the primary agent of human character, and that the races of man were separate and unequal on every level.[27] Knox – who became infamous over the Burke and Hare grave-robbing scandal of the 1820s – had considerable influence on British race thinking. His *Races of Man: A Fragment* (1850) brought together polygeny, the ineffectiveness of environment on human development, and theories about the biological destinies of Anglo-Saxons, Celts, Jews and others. He argued that it was impossible to learn anything about the origin of man by studying a society's history, and so investigators should rely on studying the biology of the races. He performed numerous dissections of cadavers so that his theory would be supported by first-hand observation. In this way 'zoological history' would show that the races were distinct, and that social history was driven by biological laws.[28]

Maintaining a more empiricist approach, British natural philosophers tended to shun theories unless based upon a preponderance of factual evidence. While human-origin theories in continental Europe were based for the most part on intangible qualities such as history, culture and linguistics, the British looked for physical evidence. During the early part of the nineteenth century British historical archaeologists were loath to consider man's ancient past because there were no texts, manuscripts or ruins to handle and examine. It was not until the mid-nineteenth century, when geological methodology became part of the search, that human origins were seriously considered.[29] By meticulously studying the stratigraphy and geological makeup of the earth, they were able to learn a great deal about how sediment was laid down, how fossils were formed, why they appeared where they did, and, most importantly, how old the rocks were. Though they often disagreed as to some of the details, methods and meanings of their discoveries, the geologists were in general agreement that the earth was far older than biblical theories could account for.[30] The decline in the 1850s of biblical geology (which supported scripture) was followed by the rise of secular geology. The initial problem tackled by British geologists was to determine if humans had coexisted with now extinct animals: if they had lived 'among the mammoths.'[31]

The British, however, were generally more interested in 'man's place in nature' than the place in nature where man first arose. They discussed how man evolved, how various human characters appeared, and man's relationship to the primates. Some tantalizing evidence of man's antiquity – a few fossils and some stone implements – were found in England. They suggested great age for the inhabitants of the British Isles, if not for their culture as well, but those finds were inconclusive at best. Unlike the Germans, the British had a longer, more secure history and sense of cultural unity, which tempered their search for human origins. While they were interested in cultural antiquity, it was often with the view towards making the British past as colorful as that of the classical world or to show how humans had first entered the British Isles. English colonial activities on the sub-continent were opening up much of India's history to them, and with this familiarity a certain contempt arose for the cultural aspect of the Jewel in the Crown. As such Asia did not present itself to the British as a cultural starting-point in quite the way it did for the Europeans.

Though not a priority, the location of the first humans was given cursory attention by some British naturalists. In his anonymously published *Vestiges of the Natural History of Creation* (1844), Robert Chambers hypothesized Asia as the cradle of man. Tracing the supposed migration routes of various 'tribes' backwards, he saw the lines converging in Asia 'about the regions of Hindostan.'[32] *Man's Place in Nature* (1863), T.H. Huxley's argument for the close relationship between the apes and man, looked at where prognathous-faced humans existed as compared to where flat-faced ones did and came to the same conclusion as Chambers.[33] Several years later, in 'On the Methods and Results of Ethnology' (1865), Huxley weighed in on human origins. In typically iconoclastic fashion he took a middle road, saying that humans undoubted began from one ancestor, but then evolved into different varieties which may or may not be different species. Arguing that there was really little direct evidence one way or the other, he said that though humans had evolved from the lower primates, they had done this 'probably, but by no means necessarily', in one location. Huxley was not completely thrilled with the idea of the Aryans either. Though he christened the hypothetical ancient Caucasians *Xanthochroi*, he said that 'of all the odd myths that have arisen in the scientific world, the Caucasian mystery … is the oddest.' With nothing more than linguistic affinities to go on, he thought Aryan theory a 'strange intellectual hocus pocus.'[34] In 'The Aryan Question' (1890) he knocked the practice of tracing the detailed routes by which the Aryans left their Asian fastness to populate the world. Still in possession of his biting wit, he said that those who created these studies did so 'with as much confidence as if they had access to the records of the office of a primitive Aryan quartermaster-general.'[35] Adrian Desmond says that Huxley wrote this article as an old man attempting to 'pinpoint the cradle of the Aryans west of the Urals.'[36]

Published the same year as Huxley's *Man's Place in Nature*, Charles Lyell's *Geological Evidence of the Antiquity of Man* (1863) argued that man was at least one hundred thousand years old, and while not tackling the question head on, Lyell did generally support the Central Asian origins of man, at least for Aryans. He doubted that the Aryans were still around or that their descendants could really be selected out from the rest of the population. In their headlong rush to conquer Europe, he surmised, they intermingled so much with the indigenous peoples that Aryans ceased to exist as a separate race.[37] Charles Darwin, after closing the *Origin of Species* (1859) with his cryptic line about man, waited until 1871 to grapple with human evolution in detail. While Lyell was working on his human-origins book, Darwin told him that 'I have thought (only vaguely) on Man.'[38] When he did get around to it, Darwin broke ranks and looked to Africa as the cradle of man, based upon the evidence apparent in the primate line. Darwin did not see conclusive evidence and argued that 'it was useless to speculate' on such topics because of the great span of time involved, and the changing conditions that must have occurred. The *Descent of Man* (1871), like *Man's Place in Nature*, was an examination of the origins of human characteristics, the importance of sexual selection in the propagation of the species, and man's relationship to the apes.[39]

The origins of light-skinned Aryans was one matter; the origins of darker-skinned people another. Complicating the situation, the apes, that darkest-skinned 'man' of all, waited in the shadows and threatened to burst upon the stage. The subject of the primates was taking up more of the discussion of human evolution. If European intellectuals were disturbed by how closely Africans, Asians and other people of color stood next to them on the ladder of life, they were more so about the proximity of the primates. There were many peoples in the world whom Europeans saw as being little different from apes and monkeys on both cultural and biological levels. The problem was to work out what the relationship, if any, these brutish people and creatures had to European man. Beginning in the late fifteenth and early sixteenth centuries, the fear of degeneracy and decaying civilization, and even decaying humanity, crept into intellectual discourse. This feeling was an offshoot of the basic Christian concept of man's fall from grace. As Europeans learned more about the world around them they saw that many diverse cultures had similarities of speech, custom and behavior. The interest in 'savage' cultures made Europeans think more about their own place in the world. Primitive people and creatures could be indicators of how far civilized men had risen from debasement, or how far they could still fall. Many European intellectuals preferred the former. Comparing the civil and the savage became a great preoccupation of philosophers and natural historians alike, particularly in England with its well-established class structure.

The connection between apes and men was not really discussed until the late seventeenth century, when systematists began working out hierarchies for living things. As science grew as an authority in the West, classifying groups of plants, animals, people, and cultures became the order of the day. The European systematizers ranked groups as superior and inferior using their own faces as the measure. These comparisons, however, could sometimes have disturbing results: some primitive cultures were surprisingly sophisticated, often as sophisticated as Europeans themselves.[40]

The most disturbing comparison was between men and apes. Early works, like William Jardine's *The Natural History of Monkeys* (1833), argued against primates being the same species as humans (as Rousseau said they were). Despite this, primates seemed too much like some African peoples, and the Africans seemed too much like Europeans. The implications were clear, yet frightening to some intellectuals. They were not happy with the idea of being so closely related to brutes. Some tried to alleviate the problem by elevating the lower orders through the concept of the 'noble savage.' Most preferred to push the brutes further away.[41]

As long as these creatures stayed caged on the proper rung of the ladder of life, they were simple curiosities. As more knowledge of their biology and behavior was discerned, keeping them in their place grew problematic. As with primitive cultures, the apes seemed more and more like, rather than unlike, humans. Their obvious similarities to humans made them objects of both curiosity and revulsion. It was too easy to see oneself in the reflection of a monkey's eyes. A close relationship with monkeys threatened Europeans' ideas of their own elevated position and reminded them of their base origins. Though ape and evolution cartoons and references in literature had been appearing in England since the 1840s – especially after the publication of *The Vestiges of Creation* – with the appearance of *The Origin of Species* the ape, and the image of the brute, was now used as a lampoon with giddy delight. In this way, the ape connection became of interest to others than natural historians or theologians. Newspaper and magazine editors 'ever fascinated with hierarchies saw the edition of the gorilla as a sensational new motif in the long-standing British obsession with social status.'[42] The fascination with brutes, both real and imagined, grew steadily. In a stream of novels, newspaper articles, learned books and cartoons – from Kingsley's *Water Babies* to Conan Doyle's Sherlock Holmes – the connection between man and ape, as well as evolution, was ridiculed, satirized and condemned. Authors as diverse as H. Rider Haggard, Joseph Conrad and Edgar Allen Poe all explored the interaction and relationship of men and monkeys. Robert Louis Stevenson's story of *Dr Jekyll and Mr Hyde* (1866) literally turned a man into a monkey.

Jekyll, the good white man, is transformed by modern science into Hyde, a low, dark, malignant beast.[43] 'Savage' races were equated with savage beasts in the growing tide of racial stereotypes. Ape imagery dealt with race, class, the spread of empire and even gender issues, as well as evolution and human origins. The ape became a metaphor of everything dark and troubling in European minds, not only about evolution, but about themselves.[44] Almost from the start these terrible monkeys, in all their guises, haunted human evolution and origin theories, and would continue to do so.

The American Scene

The answer to the unity or plurality question in America had implications beyond being an interesting intellectual puzzle. The abolition movement was beginning to pick up steam in the early decades of the nineteenth century and whether or not the Indians, and more importantly Africans, were inferior or even human held significant political import. Proving plurality could bolster those who supported slavery and the inequality of men, and thus aid the South in the growing acrimony between slave owners and abolitionists.

The unity school was typified by the work of Samuel Stanhope Smith (1751–1819) and his *Essay on the Cause of the Variety of Complexion and Figure in the Human Species* (1789). In the first American book on racial variation, Smith argued that natural causes, not separate creations, were responsible for the different races. Smith said that though God created the first people, environment and society had produced the later variations. He felt humans were plastic and thus able to change, given the right social environment.

Support for the slave cause came in the form of Samuel George Morton's *Crania Americana* (1839). Morton was obsessed with human skulls and collected over six hundred of them, which he kept on display in his home. He used his macabre collection to research skull shape and determine whether or not the races could be ranked by intelligence based on brain size. His later work, *Crania Aegyptica* (1844), argued that slavery was as ancient as man himself, and that Blacks had always been slaves. Morton pointed out that if God had created man six thousand years ago, there was not enough time for people to have evolved. Citing Egyptian art, he said that people from that period were not noticeably different from present-day humans. Therefore there was no change, and so multiple creations were necessary to account for racial diversity. In 1848, however, he seemed to backtrack when he hypothesized that the classical Egyptians were a result of Asiatics becoming somehow 'engrafted' on to the original Nile valley population.[45] Morton's student Josiah Clarke Nott (1804–73) turned the theories of *Crania Americana* into a justification for racism. As a rabid pro-Southern slavery supporter, Nott argued that science clearly showed that Blacks were of separate origin and inferior, and thus slavery was not against the natural order of things. Nott promoted the idea that the races were fixed and unchanging, and that environment had no effect upon human variation. Morton's physician friend from Philadelphia, J. Aitken Meigs, supported the collecting and studying of such material because, as he told the Academy of Natural Sciences, the skull and brain 'exhibits race-characters more striking and distinguishing than those presented by any other part of the bony system,' and thus went a long way to answering questions of human variation.[46]

After arriving in America in 1846, the legendary Swiss comparative anatomist Louis Agassiz (1807–73) entered the fray on the side of the plurality men. Agassiz could not accept unity because it suggested that people, and by implication all life, could evolve. As a man who firmly believed in divine creation, he was one of the staunchest anti-Darwinians of the period. He argued that plurality was not anti-biblical, but in fact showed that God had more control over creation than men thought. In 1855 he said the physical differences in the races were permanent, and found unity morally repugnant because it suggested that mankind was the result of incest, which is 'revolting even to the savage.'[47]

Agassiz's biographer Edward Lurie has argued that the Swiss master was not a racist, but had been co-opted by racists like Nott and his partner George Gliddon to support their political agenda. Nott and Gliddon's *Types of Mankind* (1854), based upon the work of Morton and Agassiz, helped establish plurality of creation – and racial inequality – as a hypothesis based upon objective science, not religious speculation such as the unity theory.[48] With this debate an interesting situation arose. Science in antebellum America found itself defending slavery because the plurality camp claimed its theory was based upon empirical facts whereas the abolitionists defended their theory by scripture. Therefore, as science began to free itself from the restraints of biblical literacy and concordance, it supported political agendas and slavery. In order to attain intellectual freedom, the pursuit of science was co-opted to help justify oppression.

As visions of India danced in the heads of Enlightenment and Romantic intellectuals on the continent, Americans were taking a different approach to the subject. People in the United States had no pretensions about ancient lineages and blood lines for their new nation. The Founding Fathers searched for national unity not through ancient racial history but through a democratic political institution which would link people of diverse origins into a single society. During the early nineteenth century, as the United States turned from Jeffersonian to Jacksonian philosophy and politics, questions of human origins had social and political ramifications. In addition, American science was beginning its slow ascent from the pastime of wealthy dilettantes to an organized intellectual discipline. The investigation of North America, with its seemingly endless supply of unusual and exotic geology, flora and fauna, was the perfect vehicle for naturalists to prove they were Europe's intellectual equals.[49] However, the clash between the exotic natural elements of the New World and the traditional Christian world proved a major obstacle to this pursuit. The result was a wide range of theories awkwardly assembled to explain the evidence found in America.

The first puzzle for the newly arrived Americans to unravel was the existence of those Americans who preceded them. Samuel Stanhope Smith argued that the first humans had appeared in Central Asia in a state of civilization where they were aided and instructed by God. By the time they had migrated to America they had thrown off the veneer of civilization and degenerated to a state of barbarism.[50] Samuel Mitchell disagreed with Smith, arguing that environment had little to do with the differences between the races. In addition to believing that heredity accounted for racial variation, Mitchell argued that the first Americans had come from several places at different times. He said that Plains Indians were probably descended from 'Tartars' from Asia, the Indians of the Northeast from 'Laplanders and Scandinavians,' and the people of South America from Malaysia. Mitchell also claimed that a great war had broken out between these various peoples, with the barbarous Tartar Indians victorious.[51]

As to the origins of the Indians, there seemed almost as many different theories as there were Indian groups. The contemporary literature on the origins of the Indians, during what Raymond Murray called 'the period of rash guesses,' is too vast to examine in detail here, but it is worth considering.[52] Thomas Jefferson thought the Indians had come originally from Asia, as did John Warren (1778–1856) – the first physician to perform surgery upon an anaesthetized patient, and the author of the first American book on comparative anatomy (1822). Warren also believed Blacks and Indians as intelligent as Whites, if educated.[53] In 1672 John Josselyn thought the Indians might be 'Tartars,' and that the Mohawk language might be a variant of Turkish. Searching for anything that seemed possible (or at least plausible), some early nineteenth-century theorists had suggested the Indians' descendants came from Korea and even Kublai Khan's navy.[54] Other theories of the origins of the native Americans involved the Chinese, Carthaginians, Romans and Canaanites. In fact almost every classical civilization was offered as a source.[55]

The most popular theory concerning the origins of the Indians was that they were descendants of the lost tribes of Israel.[56] It was so widely held an idea that the Puritan heretic Roger Williams, founder of Rhode Island – a state full of curiosities – thought the Indians descendants of the Jews. Many Americans used Jewish origin theories for the Indians as a justification for their mistreatment: if they were Jewish then the barbarous treatment shown them was simply the divine fulfillment of 'Noah's curse.'[57]

By the 1850s multiple creation was a widely held theory within American ethnological circles. Without an evolutionary paradigm, the American anthropologists were forced to conjure up such notions as separate creations to account for human diversity. To explain the Indians they had to look for people and languages around the world that were similar, and then figure out how those people and languages found their way to the New World. Regardless of their ideas about origins, none of these researchers saw the age of man as being very old. William Stanton argues that the early nineteenth-century American anthropologists were chronologically conservative. They did believe men in America to be ancient, but not too ancient. The Americans were attempting to operate within a more or less biblical time frame so as not to upset Christian tradition. Stanton also argues that the reliance upon separate creations in America was a result of the distrust of what they considered the thinly supported pre-Darwinian evolutionary systems of Lamarck and Chambers.[58]

Pushing the time scale for the age of man in America back further than it had been (albeit gently), these anthropologists were helping to clear the way for even more ancient age theories of man and thus open up thinking to the immense time scales that would be required for the Darwinian system. By ushering in the pluralist view of creation and beginning the break with strict biblical literalism, these men also opened the way to looking for the origins not just of the races, but of humankind itself. If one no longer believed that God created man in the Garden of Eden, then one was free to begin to speculate as to where man had originated. During the nineteenth century there were few anthropologists or natural historians not interested in human origins. Like that of their European counterparts, their work resulted in the slow, but steady secularization of these questions.

The Occult Revival

Asia as the birthplace of man also figured prominently in the thinking of the late nineteenth-century occult revival because devotees of the movement too saw it as the starting-point for their construction of a non-Christian creation story. This flowering of interest in the mystical aspects of life was a *cause célèbre* amongst middle- and upper-class artists, poets, writers and others in the West who had become disillusioned with the coldness and sterility of the modern world.[59] The Asian origins of humanity appeared not just in movements or national traditions, but was central to the thinking of some individuals as well. One of the pivotal characters of the occult revival was Madame Blavatsky. Known as HPB to her friends, Helena Petrovna Blavatsky (1831–91) was born a Russian aristocrat who spent most of her childhood living with her grandfather, a governor for the tsar. After a stormy youth she landed in New York in 1874.[60]

Immersing herself in the growing spiritualist movement, she met journalist Henry Steel Olcott (1832–1907). Olcott was covering spiritualism for a number of New York newspapers and was interested in mesmerism and the occult. The spiritualist movement began in upstate New York in the 1840s. What followed was a widespread and popular movement in which the living believed they could communicate directly with the dead. Olcott and HPB became friends and she soon initiated him into the mysteries of 'brahmavidya,' or Eastern spiritualism. They began living together in New York while attracting a circle of other occultists, 'seekers' and hangers-on around them. In 1875 they formed the Theosophical Society. At its height the Theosophical Society included the likes of Abner Doubleday and Thomas Edison, but did not last long in its original incarnation. While she may have been an occult adept, Blavatsky was not very adept at the day-to-day banalities of running an organization. Though most members drifted away (including Doubleday and Edison), a core remained to edit her first book, *Isis Unveiled* (1877).[61]

An important convert to spiritualism and member of the Theosophical Society was British naturalist Alfred Russel Wallace. Spiritualism, with its attendant seances, spirit materializations and photographs, was a rage in England by the 1870s, though it had seduced Wallace over to mystical ideas about human evolution in 1865. Disillusioned with a materialist world-view, Wallace – the co-discoverer of the mechanism of evolution along with Charles Darwin – abandoned random variation and natural selection when it came to man's intellect and morality, and fell back on non-random causes for these factors. The study of spiritualism and phrenology convinced Wallace that man's moral and spiritual dimensions were unique. Using evolutionary theory's own method against itself, he argued that spirit had no use in the fight for survival and so could not be the product of natural selection. While a few British naturalists were mildly interested in spiritualism (St George Mivart, Francis Galton and others), only Wallace embraced it wholeheartedly. He tried to get Darwin and T.H. Huxley interested, but they, after a cursory look at the phenomena, rejected it. Darwin called spiritualism 'rubbish,' and Huxley, after attending a seance, quipped that spiritualism was 'too amusing to be fair work, and too much hard work to be amusing.'[62] Huxley was trying to get metaphysics out of British science as a way of helping establish a place for himself, and others like him, in it as rationalists.[63]

With the burgeoning interest in India and things 'of the East,' Blavatsky's writings became popular. Hers was a philosophical system through which adepts could allegedly gain knowledge of transcendent reality through revelations and other occult techniques, and thus achieve spiritual salvation and elevated consciousness. These adepts were guided by masters (mahatmas) who have lived for centuries and choose certain worthy students to whom they impart wisdom and instruction. The abode of the mahatmas was, according to HPB, Central Asia.[64]

Blavatsky's most important work was *The Secret Doctrine* (1893). According to HPB, *The Secret Doctrine* was based on the supposed Eastern text known as the *Stanzas of Dzyan*.[65] She argued that evolution works through a series of cycles, or 'rounds.' The first four rounds are fire, air, water and earth, while 'ether' makes up the remaining three. During the initial four rounds life is falling from a pinnacle; during the last three it climbs back to perfection. At the end of each cycle, the next begins at a slightly higher level. Each round is divided into seven 'root races,' with the first group descending and materialistic, while the others are ascending and spiritual. The current human race is, according to this system, in the fifth root race of the fourth round. Within this scheme, Aryans are the current superior race, the Mongolians the preceding fourth race. The Mongolians, while being associated with the Atlantians of mythology, were superior to current humans, but they represented the nadir of the downward spiral. The Aryans, while technically a lesser breed, represent the beginning of the upward trend back to perfection. Blavatsky argued that the Mongolians were guilty of miscegenation and had precipitated their fall from grace by breeding 'monsters.' These monsters were still visible as vestiges of the lesser races that currently inhabit the earth.[66] HPB claimed that man had not descended from the apes, but the other way around. Apes appeared because of the in-breeding of the Mongolian/Atlantians, which created the missing link. This man–ape link will remain a mystery, however, because 'of that which has taken place, physical science can really know nothing, for such matters are outside of its sphere of investigation.'[67]

Madame Blavatsky's form of evolution was, like many evolution theories of the nineteenth century, progressive, goal-oriented, non-random, and always leading to a higher form or level of spiritual consciousness. By her own accounts, however, she did not develop this system, but was instructed in its mysteries by the 'Hidden Masters.' Her masters, Morya and Koot Hoomi, were theosophical adepts who claimed to be advanced members of the current evolutionary group, and who instructed HPB in their rites after she found her way to their monastery in Tibet. The masters themselves had masters, the remnant of the Atlantian priest-kings who (since the Atlantians fell from grace because of the sin of miscegenation with inferiors) live in the hidden city of Shamballah in the Gobi Desert of Central Asia. It was the mission of Morya and Koot Hoomi to reveal to the Aryans the greatness of their race, and the techniques that would allow them to rise even higher on the evolutionary scale.[68]

Though she has been called an anti-evolutionist – she certainly disliked Darwinian evolution, with its emphasis on natural selection and variation – Madame Blavatsky was not opposed to evolution in general. Indeed, for her system to work, evolution of some type had to occur. Master Koot Hoomi said that man had 'evoluted' in Central Asia.[69] HPB's view of evolution was a mix of science and the occult. Drawing on, and analyzing, the work of major British and European anthropologists of the day, she structured her theory to suit her world-view. Ironically, she used the work of Darwin, Huxley, Lyell and Haeckel, arguing that the latest scientific evidence and theories supported her system. In *The Secret Doctrine* she not only claimed monkeys had evolved from men, but that man was the source of all mammalian life. Using Richard Owen's archetype concept as a model, she said that 'the human type is the repository of all potential organic forms, and the central point from which these latter radiate' (683). Therefore man is even older than scientists thought, but the earlier stages of human evolution left no physical evidence except for 'astral fossil forms' (684). The cradle of current humans was the same mysterious, lost continent of Lemuria that Ernst Haeckel talked about.

Unlike Haeckel, however, Blavatsky claimed that the creatures which left Lemuria were fully formed humans (697). They were 'Africo-Atlantians' who, as they migrated into Europe, were 'evoluted' by the climate. These early Europeans were in turn ousted by the more advanced Aryans from Asia (740–741). In addition to repudiating Darwinian evolution, Blavatsky also disavowed special creation and miracles. There are, however, spiritual forces at work in the universe that give 'PRIMARY IMPULSE [her capitals] to the law of evolution … ' (731). This impulse was the 'creative power' that drives evolution. Nature is made up of 'designers' that function as localized creators, making modifications to life forms in line with the overall evolutionary cycles. She likens these localized designers to 'masons who are working under the impulse given them by the ever-to-be-unknown (on our plane) Master Mason – The One Life and Law' (732). Though she claims that there is no central plan to this system, it does have an upward progressionist tendency towards perfection. Like the Romantics, Blavatsky was enamored of India, but saw Central Asia as the abode of the Gods and the first men because of its remoteness: secret fortresses and hidden masters need a place to be secretive in after all.

In *Isis Unveiled* and *The Secret Doctrine*, Madame Blavatsky was working in a way analogous to many naturalists and scientists of the nineteenth century who sought to create an intellectual framework within which to address pressing issues they felt were neglected by Christianity. In her view, she was constructing a theory of human origins which was well reasoned and accounted for the latest scientific data – albeit washed through an occult sensibility – and that supported her view of spiritual salvation. She was, as Peter Bowler points out, marshaling an internally logical theory based on empirical evidence.[70] Though she often retreats to 'unknowable' occult processes to explain some tricky parts, her version was no more flimsy than some 'scientific' theories of the day.

The German Mystics

By the latter part of the nineteenth century, the Romanticism of Schelgel and Hegel had been revived into the New Romanticism of Eugene Diederichs, the eccentric publisher who would help spread mystical ideas throughout Germany in the early twentieth century. At the same time Ernst Haeckel's conglomeration of nature worship, mysticism and science gave 'rational' support to the notion of the 'recreation' of the German man into a new being in tune with nature.[71] Under Haeckel, the Monist League sought to arouse an awareness of the demands and importance of biology and the dangers of the same type of race decline inherent in Madame Blavatsky's story of the downfall of ancient races because of miscegenation. Germany's political upheavals and economic crises had intellectuals searching for a unifying principle and a sense of 'rootedness' in history that they felt Germans lacked as a people. The study of human origins and the search for the origins of the Aryans went a long way to providing that unifying history, and thus many German intellectuals became obsessed with their Aryan and Asian origins. George Mosse argues that this philosophical trend was a result of a German middle class fearful of slipping into the obscurity of lower-class status brought on by the economic struggles of the latter half of the century. They were afraid of losing their privileged place in the social hierarchy. Mysticism, the occult, Aryan myth, romantic ideas about Central Asian origins all seemed to prop up their social status, where modernity seemed to undermine it.[72]

The retreat into the occult and the popularity of theosophy thus had its strongest appeal in Germany. This popularity may have stemmed from the growing non-Christian view of the cosmos and human origins being formed within a Teutonic structure.[73] An offshoot of German Romanticism that first appeared in the 1870s, the Volkish movement glorified the mystical oneness of the German people, or Volk, and sought personal fulfillment and salvation in self-identification with the land. Volkish proponents rejected Christianity as not addressing the specific concerns of Nordics, and argued for a German soul directly connected to nature in a way unlike other races (especially the Jews). Volkish theory emphasized a natural spirituality that thrived on race-based imagination and creativity. The Aryans were believed to have been the most imaginative and creative people, and the current Germans, it was argued, were their direct descendants.[74]

After Madame Blavatsky's *Secret Doctrine* was translated into German in 1901, it quickly became a central work for budding German mystics. Central Asia thus appealed to a wide range of European esotericists who were interested in human origins. Occult evolution is about advancing, not only physically, but spiritually as well. Progressive levels are attained that lift the initiate above the pale of everyday existence and the lower orders to reach a superior position in the cosmos. The Atlantians and Lemurians stand in for the Neanderthals, Cro-Magnons, and other proto-humans as the stepping stones toward perfection, while the Aryans become the ultimate level of evolution. Occult adepts must pass through a series of tests and rigorous intellectual exercises to find lost knowledge of the past that, once satisfied, is rewarded with a view of how the universe is really ordered and the redemption and exclusivity that allows one to enjoy it. Therefore, spiritual evolutionary struggle makes a better spirit in the same way as physical evolutionary struggle makes a better organism.

Conclusion

Throughout the Middle Ages and Renaissance, human oddities and 'monsters' from around the world became subjects of great interest to Europeans. They played only a small tangential role in the search for human origins, but are worth mentioning within the context of this study. Though the origins of bird-headed men, people with heads in their chests, two-headed people and other human freaks were considered, they piqued the Western imagination for other reasons. As author Rosamond Purcell points out, part of the attraction of human oddities was that they could be claimed to be from far-off places inaccessible to any European and thus fanciful explanations for their existence could be put together with little chance of being questioned. Moreover, they fulfilled the psychological needs of those investigating them.[75] This phenomenon can be seen in the Asia hypothesis of human origins, particularly in the transition from India to Central Asia as the point of focus. Pioneered by the Germans, the idea of man's Asian origins quickly spread to other researchers and philosophers in Europe, Britian and America. It was adopted by all of them for different ideological reasons. The Romantics fell in love with India and saw it as the cradle of man and civilization. The later Neo-Romantics, Pan-Germans and Volkish proponents, however, did not hold colonial India in such awe because it did not fit with their views of Aryan superiority. The Neo-Romantics were uncomfortable linking themselves to the dark-skinned people of the sub-continent, and so looked in a more northerly direction. As the century progressed, and nationalistic notions took on a more important role, Asia was seen, not as the cradle of all men, but as the cradle of Aryan man. A shift occurred as the focus of human origins moved steadily away from known geographic locations and civilizations to somewhere with no (at least in the European view) pre-existing cultures. Early ideas about the Holy Land and its Semitic inhabitants gave way to exotic dreams of India, and then finally to mist-shrouded Central Asia.

This march North, and the air of mystery it offered, is a key ingredient in the story. For a land with seemingly no history, a new history could be easily created. European human-origin theorists were free to conjure fantasies to fit their particular world-view or cultural predilections. Only in Central Asia could the Aryans romp, plan their world-conquering civilization, and develop their superior intelligence and abilities in peace without the bothersome influence of Jews or dark-skinned interlopers to upset the dream. A race's or culture's ancient provenance would give it a heightened sense of tradition, superiority and, most importantly, continuity: past greatness and racial superiority meant current and future greatness for those who could best claim it.

What was needed was evidence to support the case one way or another. Fossil human remains were being discovered as early as the 1830s. In France, Jacques Boucher de Perthes discovered tools and implements in the Somme Valley that he thought were human in origin. By 1856 the first Neanderthals were discovered in the German valley that bears their name.[76] However, fossil discoveries in Europe during the early part of the nineteenth century were invariably written off as examples of *Homo Diluvinii Testis* – people killed off by the Noachian flood. The powerful head of the French scientific establishment, George Cuvier, an anti-transmutationist, saw to it that no other explanation was taken seriously there. He said that human fossils did not exist.[77] Human fossils and stone implements were also turning up in Britain. The poor condition of fossils, and disagreement about the strata in which they were found, kept the lessons they had to teach from making a definitive case one way or the other. The fossils did not seem to prove conclusively that man was ancient. The fact that so few human remains were being discovered gave pause even to staunch supporters of human antiquity. Stone tools and other artifacts being discovered could be argued against as naturally occurring if they could not be directly linked to human remains. Addressing this issue, geologist John Phillips, in his *Manual of Geology* (1855), said that human remains would be found sooner or later, if not in Britain then certainly in Central Asia, which he believed was the cradle of mankind.[78] As will be seen in later chapters, human fossils were eventually found in larger numbers and in better condition, but it would not be until the mid-twentieth century that they were able to shed any light on human origins.

By the end of the nineteenth century, while there was a kind of consensus that the first humans had appeared in Central Asia, researchers, from scientists to philosophers to mystics, were far from unified as to how they appeared, what their relationship to modern humans was, or what it meant on a cosmic level. The field was such a mess of competing belief systems that little genuine progress was made in understanding human origins. Though each theory had its own logic and support, they shared motivations in the desire to use Central Asia as part of wider ideological needs and concerns dealing with nationalism, anti-modernism, politics and spiritual salvation. The advent of the industrial age, and the disillusionment with organized Christianity and modernity, pushed many to look for a new home for the earliest people that would be free of these encumbrances, and pave the way for a new spirituality and sense of race consciousness. They searched for a source of salvation and spiritual reward, a model for social hierarchies that would bring stability to the modern world, and a way to separate themselves from the common run of humanity.

It should also be noted that many of the origin theorists mentioned here built their work around little or no physical evidence. For the philosophers, only linguistic studies and some recourse to the history of cultures could yield anything like empirical data about Asia as the cradle of man (and that was thin at best). What drove many of these thinkers was imagination. In the battle between imagination and empiricism, the former made India, then Central Asia irresistible to Romantics and mystics. To empiricists, Asia was of only mild interest because there was little or no evidence to support it as the birthplace of humanity.

None of the people discussed in this chapter ever went much beyond speculation. Central Asia would hold her secrets tightly until Henry Fairfield Osborn rallied the resources of industrial America to send someone out to talk to her. Osborn, however, had little or no direct connection to the early human-origin theorists. His intellectual inspiration for the Central Asian origins of man had other sources – which will be discussed in the following chapters. Though Osborn should not be seen as somehow fulfilling or continuing the work of the past theorizers, he did share certain affinities with them. While having an underlying foundation built upon the latest scientific discoveries about biology, geology, climatology, animal migrations and fossil studies, Osborn was drawn to Central Asia by many of the same romantic, race ideology, and even mystical notions held by Haeckel and Blavatsky.

Notes

1. John Hedley Brooke, *Science and Religion: Some Historical Perspectives* (Cambridge: Cambridge University Press, 1993), 264. For mythological elements in paleoanthropology see Misia Landau, *Narratives of Human Evolution* (New Haven: Yale University Press, 1991).
2. Frank Turner, *Between Science and Religion* (New Haven: Yale University Press, 1974).
3. See Edward Said, *Orientalism* (New York: Vintage Books, 1979), for a more thorough discussion of this topic.
4. M.H. Abrams, *Natural Supernaturalism: Tradition and Revolution in Romantic Literature* (New York: W.W. Norton and Co., 1971).
5. Claude Blanckaert, 'On the Origins of French Ethnology,' in George W. Stocking ed., *Bones, Bodies, Behavior* (Madison: University of Wisconsin Press, 1988):18–55.
6. Ibid., 131–187. For an overview of the origins of European anthropology and ethnology see Margaret T. Hodgen, *Early Anthropology in the Sixteenth and Seventeenth Centuries* (Philadelphia: University of Pennsylvania Press, 1971).
7. See Jacques Roger, *Buffon: a life in natural history*, trans. Sarah Lucille Bonnefoi (Ithaca: Cornell University Press, 1997).

8. For the complex and little discussed history of the North Pole and its role in human-origin theories see Joscelyn Godwin, *Arktos: The Polar Myth in Science, Symbolism, and Nazi Survival* (Kempton, Ill: Adventures Unlimited Press, 1996). For the work of Bailly see Godwin, 27–31. Also on Bailly see Edwin Burrows Smith, 'Jean-Sylvain Bailly, Astronomer, Mystic, Revolutionary,' *Transactions of the American Philosophical Society*, n.s., 44:4 (1954), 424–538.
9. See Stefano Pozzi and Maurizio Bossi, eds, *Romanticism in Science: Science in Europe 1750–1840* (Boston: Kluwer Academic, 1994), and John S. Haller Jr, *Outcasts From Evolution: Scientific Attitudes of Racial Inferiority, 1859–1900* (Urbana: University of Illinois Press, 1971): 4–8. Stephen Jay Gould argues that, though Blumenbach was not a racist, his system could be interpreted as being hierarchical and thus opened up the chance for racists to rank humans according to perceived worth. Stephen Jay Gould, *The Mismeasure of Man* rev. edn (New York: W.W. Norton & Co., 1996), 401–412.
10. Leon Poliakov, *The Aryan Myth* (New York: Basic Books, 1974), 101.
11. George W. Stocking, ed., *Race, Culture and Evolution* (Chicago: University of Chicago Press, 1982), 71.
12. Godwin, *Arktos*, 38–39.
13. Said, *Orientalism*, 99–139.
14. J.G. Herder, *Outlines of a Philosophy of the History of Mankind*, Vol. III, trans. T. Churchill (London: 1803): 34.
15. Peter Bowler, *Evolution: The History of an Idea* (Berkeley: University of California Press, 1983), 105–106.
16. Marvin Harris, *The Rise of Anthropological Theory* (New York: Altima Press, 1968).
17. Arthur de Gobineau, *The Moral and Intellectual Diversity of the Races* (Philadelphia: J.B. Lippincott and Co., 1856), 457.
18. Stephen Jay Gould, *Ontogeny and Phylogeny* (Cambridge, MA: Belknap Press, 1977).
19. Daniel Gasman, *Haeckel's Monism and the Birth of Fascist Ideology* (Peter Land Publishing: Bern, 1998).
20. Pat Shipman, *The Evolution of Racism* (New York: Simon and Schuster, 1994), 76–95.
21. Pietro Corsi and Paul Weinderling, 'Darwinism in Germany, France, and Italy,' in , David Kohn, ed., *The Darwinian Heritage* (Princeton: Princeton University Press, 1985): 683–729.
22. Lemuria is often associated with Atlantis in various myths about lost continents. Like Atlantis, Lemuria has been thought to have existed in different places by different writers.
23. Ernst Haeckel, *The History of Creation* (New York: Appleton, 1879). Also see Theodore McCown and Kenneth Kennedy, eds, *Climbing Man's Family Tree: A Collection of Writings on Human Phylogeny, 1699 to 1971* (Englewood Cliffs, NJ: Prentice Hall, 1972), 133–148.
24. Daniel Gasman, *The Scientific Origins of National Socialism* (New York and London: Macdonald & Co., 1971), xiii.
25. James Cowles Prichard, *The Natural History of Man* (London: H. Bailliore, 1848). Also Nancy Stepan, *The Idea of Race in Science: Great Britain 1800–1960* (Hamden, CN: Archon Books, 1982), xvi–10.
26. Adrian Desmond, *Huxley: From Devil's Disciple to Evolution's High Priest* (Reading, MA: Addison-Wesley, 1997), 320.

27. Ronald Rainger, 'Race, Politics, and Science: The Anthropological Society of London in the 1860s,' *Victorian Studies*, 22 (1978), 51–70.
28. Stepan, *The Idea of Race in Science*, 41–45. Also see Susan Collinson, 'Robert Knox, Anatomy of Race,' *History Today* (December 1990), 44–49.
29. A. Bowdoin Van Riper, *Men Among the Mammoths: Victorian science and the discovery of human prehistory* (Chicago: University of Chicago Press, 1993).
30. For more on British geology see James Secord, *Controversy in Victorian Geology: The Cambrian–Silurian Dispute* (Princeton: Princeton University Press, 1986), and Martin Rudwick, *The Great Devonian Debate* (Chicago: University of Chicago Press, 1985).
31. Van Riper, *Men Among the Mammoths*, 74–117. Also see Brooke, *Science and Religion*, 60–67.
32. Robert Chambers, *Vestiges of the Natural History of Creation and Other Writings* ed. James Secord (Chicago: University of Chicago, 1994), 294–295.
33. T.H. Huxley, *Man's Place in Nature and Other Anthropological Essays*, Westminster Edition (New York: D. Appleton & Co., n.d.); 111.
34. T.H. Huxley, 'On the Methods and Results of Ethnology,' ibid., 231–250.
35. T.H. Huxley, 'The Aryan Question,' ibid., 276.
36. Desmond, *Huxley*, 581.
37. Charles Lyell, *The Geological Evidence of the Antiquity of Man* (Philadelphia: George Childs, 1863), 454–468.
38. Charles Darwin to Charles Lyell, January 10, 1860 in Francis Darwin ed., *The Life and Letters of Charles Darwin* (New York: Basic Books Inc., 1959), Vol. 2, 59.
39. Charles Darwin, *The Descent of Man and Selection in Relation To Sex*, 2nd edn (New York: D. Appleton & Co., 1898), 158–59.
40. Londa Schienbinger, *Nature's Body: Gender in the Making of Modern Science* (Boston: Beacon Press, 1995).
41. Margaret Hodgen, *Early Anthropology in the Sixteenth and Seventeenth Centuries* (Philadelphia: University of Pennsylvania Press, 1971): 254–432.
42. For more on the popular use of irony, sarcasm and cartoons in the discussion of human-origins and primate affinities see James Paradis, 'Satire and Science in Victorian Culture,' and Douglas Lorimer, 'Science and the Secularization of Victorian Images of Race,' both in Bernard Lightman, ed., *Victorian Science in Context* (Chicago: University of Chicago Press, 1997). Quote in Paradis, p. 157.
43. Peter Raby, *Bright Paradise: Victorian Scientific Travelers* (Princeton: Princeton University Press, 1996).
44. Lorimer, 'Science and the Secularization of Victorian Images of Race,' 219.
45. Samuel George Morton, 'An Account of a Craniological Collection,' *Transactions of the American Ethnological Society*, II (New York: Bartlett and Welford, 1848), 217–222. Also see Gould, *Ontogeny and Phylogeny*, 62–104.
46. J. Aitken Meigs, *Hints to Craniographers* (Philadelphia: Merrilew & Thompson, 1858), 4.
47. Louis Agassiz, *Essays on Classification*, ed. Edward Lurie (Cambridge: Belknap Press, 1962).
48. Edward Lurie, 'Louis Agassiz and the Races of Man,' in Nathan Reingold, ed., *Science in America Science 1820* (New York: Science History Publications, 1976), 146–161.
49. For the development of science in early America see Brooke Hindle, *The Pursuit of Science in Revolutionary America 1735–1789* (Chapel Hill: Univer-

sity of North Carolina, 1956) and John C. Greene, *American Science in the Age of Jefferson* (Ames, IA: Iowa State University Press, 1984).

50. Samuel Stanhope Smith, *An Essay on the Causes of the Variety of Complexion and Figure in the Human Species* (New Brunswick: J. Simpson and Co., 1810), 32.
51. A discussion of Smith, his supporters and detractors (as well as the unity versus plurality debate) is in Greene, *American Science*, 322–328.
52. Raymond W. Murray, *Man's Unknown Ancestors* (Milwaukee: Bruce Publishing Co., 1943), 216. For a discussion of the varied theories see William Stanton, *The Leopard's Spots: scientific attitudes towards race in America 1815–1859* (Chicago: University of Chicago, 1960). For some original sources see J.H. M'culloh, *Researches in America* (1817) and *Researches Philosophical and Antiquarian, Concerning the Aboriginal History of America* (1829). Also see Richard Harlan, *Medical and Physical Researches* (1835).
53. Greene, *American Science*, 336.
54. Robert Wauchope, *Lost Tribes and Sunken Cities* (Chicago: University of Chicago Press, 1963).
55. John T. Stuart, *The North America of Antiquity* (New York: Harper and Brothers Publishers, 1880).
56. L.E. Huddelston, *The Origins of the American Indians; European concepts 1492–1729* (Austin: University of Texas Press, 1967), 34.
57. Allen, Don Cameron, *The Legend of Noah* (Urbana: University of Illinois Press, 1949), 126–127.
58. Stanton, *Leopard's Spots*, 89.
59. For an overview of this period see Ellic Howe, *The Magicians of the Golden Dawn, a Documentary History of a Magical Order 1887–1923* (Wellingborough: The Aquarian Press, 1985).
60. Robert S. Elwood, 'The American Theosophical Synthesis,' in *The Occult in America: New Historical Perspectives*, eds Kerr and Crow (Chicago: University of Illinois, 1986), 115–116. Blavatsky is one of the more unusual figures of the nineteenth century whose life has been argued over extensively, particularly as to whether she was a fraud or a genuine mystic. She has been accused of plagiarizing a good deal of her work from sources both scientific and esoteric. Besides works mentioned here see Sylvia Cranston, *H.P.B.: The Extraordinary Life and Influence of Helena Petrovna Blavatsky, Founder of the Modern Theosophical Movement* (New York: Putnam's, 1993), Peter Washington, *Madame Blavatsky's Baboon* (New York: Schocken Books, 1995), 45, and W.T.S. Thackara, 'Notes on *Madame Blavatsky's Baboon*,' *Theosophical History*, **6**:8 (October, 1997), 30–40.
61. Elwood, 'The American Theosophical Synthesis,' 116.
62. Quoted in Malcolm Jay Kottler, 'Alfred Russel Wallace, the Origin of Man, and Spiritualism,' *Isis*, **65** (June 1974), 145–192.
63. Adrian Desmond, *Archetypes and Ancestors: Paleontology in Victorian London 1850–1875* (Chicago: University of Chicago Press, 1984), 25.
64. Rosemary Ellen Guiley, *Harper's Encyclopedia of Mystical and Paranormal Experience* (New York: Harper San Francisco, 1991), 611–612.
65. Helena Petrovna Blavatsky, *The Secret Doctrine* (London: Theosophical Publishing Society, 1893). All page numbers are to the 1978 Adyar, India edition.
66. Nicholas Goodrick-Clark, *The Occult Roots of Nazism* (New York: New York University Press, 1992), 17–22.
67. Blavatsky, *Secret Doctrine*, 209.

68. Alan Boyd Kuhn, *Theosophy: A Modern Revival of Ancient Wisdom* (New York: Henry Holt and Co., 1930), 21, 198.
69. *The Mahatma Letters to A.P. Sennett* (1921), Vol. X, viii, 121.
70. Peter Bowler, *Theories of Human Evolution, a Century of Debate 1844–1944* (Baltimore: Johns Hopkins University Press, 1986), 9.
71. Gasman, *Scientific Origins*, xiii–xxv, 7, 92, 147.
72. George Mosse, *The Crisis of German Ideology* (New York: Grosset & Dunlap, 1964), 7.
73. Goodrick-Clark, *Occult Roots of Nazism*, 27–31.
74. Fritz Stern, *Politics of Cultural Despair* (Berkeley: University of California Press, 1961), and Mosse, *Crisis of German Ideology*, 31–50.
75. Rosamond Purcell, *Special Cases: Natural Anomalies and Historical Monsters* (San Francisco: Chronicle Books, 1997).
76. James Shreeve, *The Neandertal Enigma: Solving the Mystery of Modern Human-origins* (New York: William Morrow & Co. Inc., 1995), 25–35.
77. Donald Grayson, *The Establishment of Human Antiquity* (New York: Academic Press, 1983), 99.
78. John Phillips, *Manual of Geology* (London: Richard Griffith, 1855), 435–439.

CHAPTER TWO

'The Helmet of our Salvation'

On a late spring day in 1877, nineteen-year-old Henry Fairfield Osborn walked slowly around a display of fossils in a room on the campus of Princeton College. His tailored, fashionable suit gave him the air of a European aristocrat. In the hall the clicking of his shoes on the polished wood floor made the only sound in the still, dry air that smelled of the heat of the approaching summer. The fossils had been arranged by Professor Arnold Guyot to show the steady, divinely ordered progression of life on earth. To Guyot the lesson seemed so clear and elegant that its truth was self-evident. The young Osborn agreed. He had recently become interested in natural history as a result of attending Guyot's lectures. So inspired, he was about to leave the bucolic splendor of Princeton to hunt fossils in the wilds of the Western territories. Having finished his inspection of the display, he turned and quickly strode down the hall and into the open air on his way to wrestle with eternity.

Henry Fairfield Osborn was born in 1857 to a world of plutocratic privilege. Despite this his parents continually strove to teach him the importance of struggling to achieve, instead of resting on advantages given him by his birth and social position. His formal schooling was first rate; he studied with some of the leading scientists and educators of the nineteenth century. His father taught him to struggle for reward in this life while his mother taught him to struggle towards success in the afterlife. The idea of actively working to advance oneself in both the physical and spiritual realms became an underlying motif in his personal and professional life and led him to search Central Asia for the origins of man as well as of himself.

The Osborn Family

Henry Fairfield Osborn's parents, William Henry Osborn (1824–94) and Virginia Reed Sturges (1830–1902), came from old Yankee families. Leaving home when he was thirteen years old, William Osborn found employment as a clerk for the East India company of Peel and Hubbell, and so impressed his superiors that by the age of sixteen he was representing the company in its Manila office. By his mid-twenties he was doing well enough to open his own company, and he became wealthy. Unwilling to rest on these accomplishments and retire, he returned to America in 1853 where he met an old business acquaintance, Jonathan Sturges, and his daughter Virginia.[1] Immediately smitten with her, William was forced to stand by and watch as Virginia left for an extended tour of Europe. Refusing to let this obstacle impede his progress, he set out to overcome the dilemma, and began courting her by mail as she tripped across the Old World.

Virginia loved England, which, because she had read so many Dickens novels, felt comfortable and familiar. In contrast, France seemed dreary and cold. There she made up for the bad weather by admiring beautiful art during the day and visiting beautiful people at night. One evening she dined with Alexis de Tocqueville and his wife. Madame de Tocqueville shocked Virginia by regaling her with tales of French marriage rituals. Miss Sturges believed that marriage should be based upon love and respect and saw the French way as empty, hollow, and an affront to much she held dear. She also had a strong sense of the divine acting in the world. While in France she wrote to her parents that 'I think more everyday of my life how providence overalls all things for our good.'[2] Whereas she may have found French courtship too risqué, William's letters made her giggle with girlish delight because they were 'exceedingly spicy and pungent.'[3] They had grown so intimate through their correspondence that they wed shortly after her return. Though the marriage ceremony left her ill, a trip to Lynn, Massachusetts refreshed her enough to begin building the family that made its home at 32 Park Avenue, New York City.[4]

Virginia Osborn engaged in the social life of New York, spending time with the Morgans and hobnobbing with the Vanderbilts and other families of reputation. She and William were often guests at Biltmore, the Vanderbilt estate, and vacationed on Nantucket. An art lover, Virginia put together a considerable collection. Some of her most prized pieces were works by her friend Frederick Edwin Church of the Hudson River School.

Virginia's father, Jonathan Sturges, was one of the founders of the Illinois Central Railroad. He was considered a good businessman and philanthropist, 'never forgetting his obligations to the poor, the sick, and the crippled.'[5] As a 'Christian full of zeal and benevolence' he helped found the Academy of Fine Arts which later became the Metropolitan Museum of Art, and assisted in the war on the corrupt 'Boss' Tweed and his scandalous Tammany Hall political machine.[6] Following the marriage, Sturges brought William Osborn to manage the New York offices of the faltering Illinois Central. William's imagination and aggressive style helped reorganize and streamline the company, and put it on a more solid financial footing. In response to this success, he was made president later in 1853.

Having put the New York office in order, Osborn traveled West to take hands-on control of operations in the field. He was part of a wave of Eastern entrepreneurs who went West to take advantage of the growing railroad opportunities that existed there. He brought shrewd Yankee business acumen to what at the time was a failing venture to open the country's interior to mechanized travel, and saved it. He was particularly adept at lining up overseas financial backing from English investors, and keeping the big unwieldy company running smoothly. He was the classic gilded-age 'captain of industry' who took the reins and guided the company according to his own vision. Osborn was also an innovator in the use of advertising and promotion to further the company's interests.

As the railroads spread West in the 1850s and 1860s, the United States government gave or sold extensive tracts of land to the railroads in order to facilitate national expansion. The Illinois Central recruited people from the Atlantic seaboard and directly from Europe to come and live on the lands that surrounded the right of way. This strategy helped fill up the country, eased congestion in the Eastern cities, and provided a ready supply of clients – as renters, land owners and shippers – for the railroad. Osborn sent agents to Germany, Norway and Sweden to sign people up and send them to Illinois as homesteaders. This tactic worked so well that by 1863 a steady stream of immigrants was heading to the American Midwest. Critics argued that the railroad companies were cynically taking advantage of the unwary peasants by sending them to a future that was considerably different from what they had been told awaited them. In response, Osborn insisted his European agents be honest and make no claims for the conditions in Illinois that did not actually exist. The company prospered and so did William Henry Osborn.[7]

As a way of getting clear of the hustle and bustle of his grueling work schedule, William and Virginia loved to venture out to the New York countryside of the Hudson Valley, where they bought a vacation cottage near Buttermilk Falls. Financier J.P. Morgan, one of William's business associates and friends, spent extended periods with them during the summers of 1858 and 1859. He whiled away long summer days relaxing with the Osborns and cavorting on the parlor rug amongst the Victorian bric-à-brac with one-year-old Henry Fairfield.[8]

In 1861 J.P. Morgan married Virginia's sister Amelia – known as Memie – and became Henry Fairfield's uncle and future benefactor. Artist Frederick Edwin Church also spent the summer of 1858 with the Osborns. The entire party went to visit the Niagara Falls. Church was so taken with the Falls that he was inspired to do the painting which became of one of his most famous works.[9] In 1859 the Osborn family moved into their new home at Garrisons, New York, on the Hudson River not far from the United States Military Academy at West Point. Christened 'Wing-and-Wing,' the estate became the family seat and remained so well into the twentieth century (William soon began work on a grander building just up the road, called 'Castle Rock' and finished in 1881). It was a splendid house on a beautiful expanse of land that had originally been purchased from local Indians in the 1600s on a site that once held a historic church and later a revolutionary war hospital.[10] 'The mountains are not too high to climb and the views from their summit is glorious.'[11] Church built his ornate Arab fantasy house, 'Olana,' next door.

The bucolic setting was young Henry Fairfield's playground and spiritual home to which he would return over and over during his entire life. Born two years earlier on August 8, 1857 in Fairfield, Connecticut, Henry Fairfield, affectionately known as 'Fairfie,' was a happy, gurgly, precocious child. He and his older sister Virginia filled 'Wing-and-Wing' and the family's Manhattan town house with childhood energy as they enjoyed the summers and toy-filled Christmases. Little Virginia loved to sit for hours with a pile of books and read out loud. Henry Fairfield's serious interest in the natural sciences did not manifest until his college days, but by the age of three he was already showing signs of professorial bearing and a desire to teach others. Clinging to the banister, he would trundle down and recite various prayers and bits of spiritual insight to whatever audience was at hand. Dutifully listening to her son's sermonizing, Virginia said 'everyday he talks up something new.'[12]

As he grew, Fairfie engaged in the usual boyish outdoor pursuits at 'Wing-and-Wing.' He loved to fish and play in the woods surrounding the estate, as well as swim in the Hudson River. Often accompanying him was his younger brother Frederick, who would bring over his friend, little Theodore Roosevelt – the future president of the United States. Amongst his frolics Fairfie practiced horticulture by planting a peanut garden with his youngest brother William, four years his junior. Willie and Fairfie had a close relationship. Though they lived in New York, few of the Osborn children were actually born there. As a wit would later put it, 'William Church was born in Chicago where an Osborn was only a man. Henry Fairfield was born in rural Fairfield, Conn., where an Osborn was decidedly an institution.'[13] William went on to a successful career as a lawyer, judge and politician of scrupulous honesty and fastidiousness, but as a youth he took a less serious approach to life. During their flirtation with being peanut farmers, Fairfield planted his seeds in neat, tidy, organized rows, while William simply scattered his and let them grow where they fell.[14] Life at 'Wing-and-Wing' in the 1870s was a gilded-age fairytale. Years later, in the midst of the turmoil of the twentieth century, Henry Fairfield said that he would 'gladly go back [to that time] with its stern convictions of what is right and what is wrong, with its pioneer spirit of Christian education and civilization.'[15]

While the country tore itself to pieces in the Civil War, and then tried to reconstruct itself, the family traveled extensively. They summered at Garrisons and wintered in Manhattan at 32 Park Avenue. Like many in the North, including Abraham Lincoln, Virginia Osborn saw the war not necessarily as a conflict to end slavery, but as a struggle to save the Union and uphold law and order. For her the North's actions seemed 'inspired by God.'[16] She argued over the war with J.P. Morgan, asserting that the Constitution was a document worth preserving, while he was not so sure.[17] She was thrilled to watch New York's Seventh Regiment march off to the war, but was apprehensive since her brother Edward was an officer in that unit. With the war raging, William moved the family to Chicago while he expanded the Illinois Central into the South. Fairfie caused a stir amongst their neighbors on Michigan Avenue in 1863 when he hung a captured Rebel flag in the upstairs window, prompting questions about whether the trophy was a sign of Union support or Southern sympathy.[18]

Before the war William Henry was partnered with George McClellan (one of their lawyers was the young Abraham Lincoln). Osborn did not think highly of McClellan and when, as general of the Union army he ran for president against Lincoln, Osborn scoffed at his old partner's chances.[19] Though they constituted a patriotic household, William and Virginia tried to keep the horrors of war at arm's length; even the infamous New York draft riots just downtown seemed a world away. They could not keep tragedy away completely. Virginia's sister Memie did not live to see the end of the war. A sickly young woman, she died in 1864.

Following the war, when business kept William in the States, Fairfie took charge. By then in his early teens, he acted as head of the family, kept control of his siblings and expenses, and made arrangement for their travel accommodations.[20] Yet the ability to run the family did not come easy to Fairfield. He went through a period of painful shyness that his father sought to cure by having the boy express himself the way he did as a child. The elder Osborn felt that anything worth having could be achieved only by overcoming the difficulties associated with it. William insisted that the boy begin publishing a newspaper called *The Boy's Journal*, which was circulated to all and sundry.[21] The rigors of writing and publishing a newspaper would be the effort needed to help him conquer his awkwardness.

Putting out thirteen editions in 1871 seemed to have the desired effect, since by later that year he was leading the family around Europe, dealing with hotel managers and train conductors with relative ease. It was Fairfield's first taste of the positive power of struggle. He sent a copy of *The Boy's Journal* to his uncle then in Egypt on business. In reply, his uncle described the sensuous delights of Cairo in such detail that William Henry playfully chided the boy, saying that now he had heard of the place he would soon run off to North Africa with visions in his head of harems and Eastern potentates.[22]

When William Sr did finally join the family on a trip to Germany in 1871, he was preoccupied with trying to find a doctor to treat his chronic 'aching limbs.' His workaholic ways were catching up with him and leaving him in excruciating pain. A Frankfurt doctor told Virginia that 'all Mr Osborn's suffering arose from a highly strung nervous organization.'[23] When he complained about a bad rash in addition to the problem with his limbs, the doctor said that he had probably 'inherited' a weakness in his left side from his parents. Not even the recuperative baths of England and Europe made him feel better for long. The problem plagued him the rest of his life, and in his last years kept him bed-ridden and suffering from fits of dementia.

Virginia Osborn

Henry Fairfield Osborn found the role model for his outward, personal behavior and business-like approach to life in his father, but his religious convictions were inspired by his mother. She received a formal education and was a voracious reader and thinker. She kept a list of books she read, which included *Wuthering Heights* and Hawthorne's *Blythdale Romance*, but also weightier tomes like James McCosh's theological works, a biography of Galileo, Plato's *Dialogues*, and Hugh Miller's geological treatise, *Old Red Sandstone*.[24] In 1844 she studied works on the origins of the races in Europe and even speculated on how those races affected the course of English literature.[25] In the 1850s she was already asking herself questions about salvation and the afterlife. 'It is hard to believe that the young and happy can die. I say to myself I may too die.'[26] The year Henry Fairfield was born she saw connections between nature and the Almighty, telling her father that 'our Savior's teachings are those beautiful associations of natural objects with our inward and future life.'[27] Virginia Osborn was raised in a Presbyterian environment which could not embrace strict Calvinism and its pessimistic view of predestination and the debasement of mankind. Years later Henry Fairfield Osborn recalled that the Sturges followed the 'lighter, happier and truly Christian side of religion, chiefly exemplified by good works, benevolence, and patriotism.'[28] Virginia came from a tradition that was more Pelagian in that she felt good works, personal effort and exercise of free will could all contribute to the achievement of salvation. Being born to wealth, whether spiritual or economic, was not enough, one had to go further. 'We would be poor, weak, delicate plants if only visited by prosperity,' she said.[29] Struggle was an important element, for without it no advance could be achieved. Looking at the life of an aunt who had given up her comfortable middle-class existence to follow her husband into missionary work, and yet bloomed into a capable woman, she marveled that here was proof that adversity and struggle bred 'brilliance.' Virginia was part of a wave of people across antebellum America, not usually thought of as intellectuals, who applied Enlightenment ideals to the study of religion as a way of creating new ways of practicing their faith. These ideals, alongside the parallel rise of Romanticism, included reason, the collecting of empirical facts, and a less dogmatic sensibility.[30]

In addition to her philosophizing, she took an active role in life. Wishing to improve what she saw as the deplorable state of nursing in New York,

Fig. 2 Virginia Osborn. It was 'Mother Osborn's' gentle Christian teachings that became the framework of her son's theory of evolution and set so many wheels in motion.

Neg. No. 104877, courtesy Department of Library Services, American Museum of Natural History.

Virginia began a nursing school which she hoped could be ranked amongst those of England. Her Bellevue Training School for Nurses, opened in 1873 and based upon the ideas of Florence Nightingale, was hailed as a model for the rest of the country. Virginia Osborn put up most of the money and took a hand in its organization. 'We must,' she argued, 'have more trained nurses in America and we must have them not only for the rich, but for the poor as well.'[31] To staff the school, she personally paid the salary of Sister Helen, an Anglican nun and trained nurse, brought from London.[32]

Virginia's preference for a simple uncluttered religiosity was expressed in her unpublished novel. 'The True Cross' is a dialogue between an old man and a 'traveler' who meet at a crossroads and begin discussing the meaning of life and the nature of religion as exemplified by the cross upon which Christ was crucified.[33] The traveler argues that few understand the meaning of the cross, and why Jesus died to show us the way to redemption, because they cloud their minds with ritual and theological infighting and forget the central message. The dialogue is an examination of what it means to be Christian. When the old man finally asks the traveler directly whether he is a Catholic or Protestant, he replies that the only true belief system is the belief in Jesus. 'By that alone,' he proclaims, 'am I called Christian.' Virginia saw good in most denominations, but one's personal approach to faith was more important than the slavish aping of organized belief. Though an avid church attendee, she sought her own way to spiritual communion. Virginia eschewed organized religious dogma, feeling that no one faith had control of the truth. Imparting her ideas to her children, Virginia's actions were part of what Barbara Welter calls the feminization of American religion. As mother of the Osborn clan she took control of how her family learned and worshipped.[34]

Virginia Osborn believed that people needed no religious hierarchy to lead them. She felt that one's consciousness was a distinct entity, with abilities and intuition to understand nature and the universe. She felt, for example, that there was 'little talent in the English church' though she personally admired the famous Reverend Pusey.[35] Once, in Heidelberg, Germany, she planned to attend an Anglican service, but changed her mind when she discovered it was of the 'High' church variety. Instead she went to a nearby Scots Presbyterian service. She had an aversion to the showy, Roman Catholic type of service with its heavy symbolism, ritual and priesthood, and leaned towards the Scottish way because of its simplicity of worship and uncluttered theology. Though she felt comfortable with the notions of providence and aspects of predestination, the only fixed creed for Virginia was Jesus. She believed divine providence worked daily in people's lives and helped guide those who were in tune with it.[36]

Faced with any type of depression or doubt by anyone in her family, Virginia would brace them up by telling them to put their faith in Jesus and to exert more effort to overcome life's hardships. She was constantly lecturing people in this fashion. Very few of the mountain of surviving letters she wrote do not contain at least one reference to Jesus. She could quote Bible passages at will and always knew the right phrase for any situation. In 1872 she wrote to her younger brother Henry about the importance of a spiritual life. He had been deeply depressed for a good part of his twenty-six years and she desperately tried to cheer him. She explained that he was 'starving [his] soul' by not allowing God to bring him peace, that pleasure and good times were empty ways to find solace, and that 'the struggle goes on 'till it is no wonder you are tired of it all.'[37] For her, life was a combination of divine guidance and personal action. Providence could only do so much; for true salvation the individual must actively embrace and use what God had given. She instilled these feelings in her children and taught them that they had a responsibility not to be carried along by others, but to make their own way.

'Mother Osborn,' as she was later known, was the matriarch of the family. In her perfectly kept Victorian attire she thought long and deep about what made a proper Christian life and drilled these ideas into her children. She particularly doted on Fairfie, calling him 'the child of her prayers' and imparted to him the basic religious convictions that drove him.[38] Mother Osborn's teachings became part of the inner framework upon which Henry Fairfield Osborn attached other ideas, particularly those of social order and the origin of man.

Princeton

At Garrisons the Osborn children were wrapped in the warm blanket of the family, protected from the outside and nurtured between the hard-working, forthright and commanding presence of William, and the loving, spiritual teachings of Virginia. Both parents stressed the need for their children to live useful, original lives, to stand up and be someone who made a difference. They also insisted that their children receive a good education. Fairfield attended the Columbia Grammar School and then the Lyons Collegiate Institute of New York City, both of which catered to the children of the gilded-age wealthy. Whereas the tradition was for Lyons graduates to go to Yale, in 1873, at the age of sixteen, Fairfie entered the College of New Jersey (later Princeton) and took his seat in the class of 1877. His maternal great uncle, the Reverend Ebenezer Pemberton (1704–77) had been one of the three founders of the school before the Revolution.[39] Initially begun as a bastion of 'New Light' thinking during the Great Awakening, the curriculum at Princeton attempted to bridge the gap between the cold intellectualism of the 'Old Lights' and the emotionalism of the radical evangelicals. Among the school's presidents were Jonathan Edwards, one of the leaders of the Awakening, and Samuel Stanhope Smith, who wrote pioneering works on racial variation, the origins of the Indians, human hereditary plasticity, and the Asian origins of man.

In the all-male world of constant ribbing, put-downs, in-fighting and testing of each other's characters, Henry seemed initially unsuited to college life.

His short cropped hair, high collar and button-down jacket did not make for an imposing sight. In response to his being 'almost girlish in appearance' he was immediately dubbed 'Polly' Osborn.[40] As a freshman he took the basic courses, including Greek, Latin and Algebra, and was a solid though not brilliant student, eventually graduating in the middle of his class.[41] He liked extra-curricular activities, particularly concerts featuring the works of Mendelssohn. He was a boating enthusiast and a member of the sculling team, and an avid football fan.[42] Though he never played the game himself, he was a major booster of the school's gridiron and attended every Princeton/Yale football game until his death.[43] He also attended Bible classes and church services regularly. He made friends easily and during the early days lived off campus in a 'low and musty' room which was later replaced with a brighter one he shared with a roommate, festively decorated with boating memorabilia and other artifacts of young manhood. Thus occupied during his freshman year, he humorously told his father that he had not yet run off 'to the flesh pots of Egypt,' though he thought he might after Thanksgiving. His father was so preoccupied with railroad finances that Fairfie, now known as Harry, had to ask his sister to remind the patriarch to send him money to live on.[44]

Harry filled his days with Ovid and algebra as well as the various student high jinks that were rife on campus, including the 'Whig Hall' debating society. While he may have looked weak, 'Polly' soon showed his mettle. During a 'cane spree,' where platoons of underclassmen attacked each other in ancient Greek hoplite formation with bamboo and wood canes, frantically trying to beat the other into submission, Osborn was the last man standing. The next day he was congratulated for his resilience and fortitude by upperclassman Charles Scribner Jr, who had witnessed the mêlée from the sidelines. The future book publisher was astounded by Harry's performance and a life-long friendship began.[45]

His mother sent encouraging letters to her boy, which were designed to keep him on the straight and narrow and to side with the right element now that he was not under her direct control. She told him to be sure to read history at least one hour a day, and to take his place at the head of the class. 'Make up your mind,' she told him, 'to lead in college … not be led by the more worthless half of class.'[46] He had, she told him, 'a responsibility to work hard' and go where his 'conscience and talents may lead.'[47] She also sent him money, not for himself, but to send to charity. 'I want all my children to form the habit of giving. It is never acquired late in life.'[48] Seeing an improvement in his scholastic ability, she pronounced that the change of environment – particularly the cold room he was living in – had stimulated his mind to greater success. His sufferings in those conditions, she told him, were a benefit in the same way Jesus's sufferings were.[49] She continually reminded him that he must work hard and accept Jesus as his savior. Doing so was 'the helmet of our salvation.'[50] To Virginia the search for salvation and belief in the recuperative powers of Jesus were like a suit of armor protecting the soul from harm. The crucial part of that armor was the helmet. But one had to choose to wear it; otherwise it was useless and salvation unobtainable, no matter what other advantages one might have.

His fortitude, strength and sense of charity were soon tested in other ways. As Henry Fairfield Osborn began his academic career tragedy struck quickly and savagely. On November 23, 1873 the steamer *Ville du Havre* sank in the Atlantic, taking over 200 lives with her – including several of Osborn's perpetually Europe-bound aunts and uncles, and the entire family of one of his Princeton classmates.[51] Then in 1875 both his younger brother Frederick and his older sister Virginia died unexpectedly: Frederick in a swimming accident in the Hudson River and Virginia of an illness while the family was in Paris. In telling Fairfield of his sister's death, his mother said that 'her twenty year sojourn is at an end.' But there was a bright spot to it all, because 'God has called us as a family to a special dedication of ourselves to his service.'[52] The tragedy was God's way of testing the family's ability to deal with adversity. Mother Osborn suggested that if the departed children still watched over the family in spirit form, 'as I firmly believe they do,' then their deaths had purpose and would benefit them all.[53] The deaths shocked and hurt Fairfield, but he buried his feelings and resolved to go on.

A Natural Historian

Spending his early years at Princeton as a dandy and college wag, it was not until his junior year that the sciences became a major interest for Harry Osborn. Reading Dana's *Manual of Geology*, and Tyndall's *Glaciers of the Alps*, he developed the urge to go camping. He had taken the geology course of Arnold Guyot and found it enjoyable.[54] Arnold Guyot (1807–84) was already a naturalist of note when he came to Princeton. Born outside Neuchâtel, Switzerland, he went to the university there and later studied for the ministry in Berlin, where he met Louis Agassiz. Having a lifelong interest in the natural sciences, he abandoned the ministry and instead did his doctorate on the geology of lakes. Venturing to Paris, he encountered Agassiz again and was encouraged by him to study the dynamics of glaciers. In 1839 Guyot was appointed professor of history and physical geography at Neuchâtel but was forced to flee in 1848 when a revolution broke out. Following Agassiz to the United States, Guyot lectured at the Lowell Institute in Boston. These talks were then published as *Earth and Man* (1849) and represent his synthetic view that divine law brought together Genesis, the evolution of the earth, and the history of man into one great story of life.[55] In 1854 he was hired by Princeton to become professor of geography and geology.

At Princeton Guyot introduced teaching innovations such as the use of maps and visual aids, and suggested his students get out into nature and observe it first hand (though he never took students out as part of the class). He was well liked by the student body, though they often tittered at his labored English. His religious views were so strong he was considered 'Christocentric,' and he saw intellectual pursuits as a homage to the glory of God, nature as a product of divine intervention, and defended the cosmogony of the Bible – though not its detailed literalness.[56] He was also interested in the natural history of Asia and translated several German works on the subject into French. In his lectures Guyot stressed the link between history and science; in his view both were examples of God's working in the universe. Quickly setting up a museum at Princeton to assist the students in learning their lessons, he used a style which emphasized 'synoptic' teaching: the treasures of the earth were laid out to show nature as an open book, illustrating the sequential history of life from its earliest beginnings to the advent of man – God's most advanced organism.[57]

Guyot would start every semester by telling his students that geology was the study of the physical changes of the earth as well as the life forms on it, and that God had made the earth incrementally. Guyot believed in slow geological change which was evidenced in the geology and strata of the planet. He regarded the earth as a living organism and believed that it developed 'progressively' – slowly and according to God's laws. Students learned that all this was supported by direct observation of nature. Facts and empirical evidence were important to the Swiss naturalist, for without facts to back them up theories were no more than conjecture, and scientists ought avoid such things. He insisted that though the world might be 'going on as it did in the age of Moses, David, and Mathusila [*sic*],' there was no way of knowing for sure unless evidence was found to support the notion.[58]

Like philosopher Henry David Thoreau, Guyot saw nature as a sublime entity. He had been going on regular trips to the Catskill Mountains since 1862 to observe and collect specimens for use in his research. He loved traveling through the spectacular scenery, taking notes in raptured awe at how the forest and mountain grandeur showed the very hand of God. *Creation* (1884), his last work, argued that the Book of Revelation and the book of nature were not incompatible: both were witness to Creation. Everything in nature, he felt, proved the sure steady hand of God in a way that was self-evident.

His interest in science growing, Harry Osborn would occasionally spend time with Princeton's geology don. As Charles Darwin took walks with his Cambridge professor Henslow, Osborn and Guyot would occasionally take strolls around the campus paths and gardens, discussing and admiring nature. During one of these walks Guyot stooped down and picked up a fallen leaf. As the late fall sunlight filtered through the trees he stared at it intensely.

Then the old man said, in a reverie as much to himself as to his young protégé, that he did not understand how anyone could look at a leaf and not believe that 'it proceeds from a divine intelligence.'[59]

Though Guyot and Osborn shared an intellectual bond, the strongest voice in Osborn's college career was that of the school's president, James McCosh (1811–94). Princeton's reform-minded leader was one of a line of mentors who permeated Osborn's life and helped forge his views on intellectual problems and how to deal with them. Born in Scotland, McCosh studied at Glasgow and Edinburgh in the 1820s and became a minister. He began his ministerial duties in the midst of a split in the Scottish Presbyterian Church between moderates (those following the philosophy of the eighteenth-century Scottish Enlightenment which was pushing the Church away from its strict Calvinist origins) and the evangelicals who opposed them. McCosh came down on the side of the evangelical camp, and supported that group's attempt to limit the workings of the state and wealthy élites in church affairs. Unable to reconcile their differences, the evangelicals split from the Church to form the Free Church of Scotland. McCosh was not, however, a dogmatic evangelical. He attempted to bridge the gap between the cold intellectualism of the moderates and the emotionalism of the evangelicals in much the same way that Jonathan Edwards tried during America's Great Awakening. McCosh's first book, *The Method of Divine Government* (1850), attempted, in the style of William Paley, to show that the complexity of nature was a sign of its divine inspiration – though with an emphasis upon the human mind, its emotions, and will as evidence. For the Scotsman the workings of the mind showed that God ran the world in an orderly fashion according to his laws. The success of the book led to his appointment to the Queen's University, Belfast, Ireland in 1852.[60]

As a growing force in Presbyterianism, McCosh toured the United States in 1866 and became involved in the tension there as well. He defended the Gospels from the perceived attack made upon them in Ernest Renan's *Life of Jesus*, and argued the right of Christian congregations to be autonomous.[61] In 1868 he accepted an invitation to become president of the College of New Jersey. At Princeton he was viewed as something of a reformer, though not a radical one. While Harvard's president Charles Eliot was overhauling his curriculum, eliminating required classes and downplaying the classics, McCosh held fast. Not wanting to throw the baby out with the bathwater, McCosh argued that there were eternal values to be learned in a classical education, though he did revamp the science department by introducing the teaching of evolution, and allowed for the taking of electives by upperclassmen.[62]

In his teachings at Princeton, one of McCosh's major areas of interest was the human mind and its workings. He opposed the notion of the relativity of knowledge which had begun to gain vogue within some intellectual circles.

He argued that there were absolutes of truth and meaning because humans could conceive of such things intuitively. McCosh believed in the transcendent power of human intuition, and opposed the *tabula rasa* concept. In *Intuitions of the Mind Inductively Considered* (1860) he argued that the human mind had certain fundamental ideas and principles built into it. Intuition was the process by which these principles were discovered by the individual. This 'primitive cognition' was the foundation upon which all common truths rest and was the basis for all religion. He thought that intuition plus experience was how people learned about nature and thus the divine laws by which nature operates. Like Arnold Guyot, McCosh argued that all knowledge, even that of religion, had to be based upon factual evidence or it was mere speculation. McCosh resisted being labeled an Augustinian or even a Calvinist, and, like Virginia Osborn, believed that people should come to their own conclusion about a personal relationship with God, and avoid falling into the trap of mindless dogmatism. He did not want 'any mere man his master.'[63]

In addition to his views on intuition, James McCosh was also a pioneer in the teaching of evolution in America. Meeting opposition for his outlook from Princeton Theological Seminary head Reverend Charles Hodge, McCosh persisted. Hodge argued that Darwinism was a rejection of a God-created universe. He equated denial of design with a denial of God. Like many in the United States who did accept evolution, McCosh was not a pure Darwinian. As far back as 1856 he elaborated his ideas about the design of life. In *Typical Forms and Special Ends in Creation*, he borrowed British naturalist Richard Owen's transcendental morphology and archetypes to argue God's handiwork, claiming there was unity and order in the universe. When *The Origin of Species* appeared, however, McCosh saw the importance of Darwin's approach, dropped his previous position, and adopted a teleological form of transmutation.[64] He accepted the fact of evolutionary change, but argued it was a divine law guided by the Almighty. He also argued that students were in no danger from exploring Darwin's work and insisted that exposure to it reinforced Christian belief. If evolution were 'properly' understood, the Scotsman said, it would be seen as the mechanism of God for creating change in the world and not as antagonistic to revealed religion. While he was comfortable with the evolution of man's physical form, McCosh had reservations about the less tangible parts. He believed morality, instinct and intelligence could not be accounted for by materialist processes. Whether there was some vital principle or internal mechanism responsible for these characteristics he was not sure, but said that science might be able to find out. We may never know anything about these elements, he conceded, except that they are unique and probably divine. Accepting what seemed an inevitable outcome, he held that man may or may not have evolved from brutes, but his soul was of higher origin.

McCosh had no time for unsubstantiated notions, either scientific or religious. He opposed atheistic evolutionary systems that attempted to take God out of the picture, but was equally wary of those who leveled blanket condemnation at science. He warned against the 'undiscriminating denunciation of evolution from so many pulpits' because it 'might drive … thoughtful young men to infidelity because they clearly saw development everywhere in nature [yet] were … told by their advisors that they could not believe in evolution and be good Christians.'[65] As he did with Presbyterian schisms, McCosh tried to walk a middle way between science and religion to bring the two together, not push them apart. Darwin's book caused great debate and turmoil in American intellectual circles. Many natural theologians came to accept it, but wanted somehow to keep the divine element Darwinism seemed to oust from the process. It is unclear when Osborn read *The Origin of Species*, but he must have read it, along with other popular works on evolution, in his last two years at Princeton. What is clear is that while he certainly embraced the central role of an organism's struggle for existence in species formation that Darwin advocated, he was far less enthusiastic about natural selection. Indeed, it would be many years before he could give natural selection any place in the evolutionary scheme of things, and never accepted it as the central force of organic change.

James McCosh imparted his ideas to his students as a professor of philosophy and psychology. In his philosophy classes students learned about the Greeks, Spinoza, Kant and others, as well as Scottish Common Sense philosophy. He asked his students, among them Henry Fairfield Osborn, to consider what truth was. He said that though it was difficult to define strictly, it could be said that a characteristic of truth was that 'it has always reference to some great standard.' Truth must be 'consistent with what is right.' He took evolutionists who thought that mind was a result of material process to task when he told his students that 'materialism is absurd in saying there is no such thing as mind.' Giving the devil his due, however, he warned that 'the philosopher who believes that everything is mental and nothing material, takes good care in going downstairs not to slip and hit his head against a hard idea.'[66]

Through their lectures and mentoring both Guyot and McCosh taught Osborn to accept religion and science as non-conflicting subjects. Building on what he learned in their classes, he wrote an undergraduate essay answering the question as to why God allowed evil in the world. Osborn wrote that God made salvation obtainable only through 'increasing effort' and constant struggle.[67]

Towards the end of his junior year Henry Fairfield Osborn sat on a canal bank in Princeton lazing with a pair of classmates, Francis Speir and William

'Wick' Berryman Scott, in the hot spring sun. The three had taken a swim and were now lying in the grass preparing for final exams by wrestling with Paley's *Natural Theology*. Osborn and Scott had become friends while in Guyot's and McCosh's classes. Roughly the same age, they were not initially close. Scott, because of his nomadic upbringing, had a heightened air of maturity which generated distrust amongst his peers and led to his being labeled a 'facultyite' – someone too closely linked with the authorities. As such he had few friends until his re-acquaintance with Harry Osborn in their junior year. During a lull as they stared up at the clouds, Wick Scott mentioned that he had recently read an article in *Harper's Weekly* about a fossil-hunting expedition out West taken by some enterprising Yale students. He absentmindedly suggested they do the same, and to his surprise both jumped at the idea.[68]

As it does for many college students, the end of their junior year brought on the first vague dawning of the idea that the carefree college days were soon to give way to their futures. Osborn was resigned to following his father into the railroad business, and Scott had plans for medical school at the University of Pennsylvania, followed by a country practice. While both were going to their fates willingly, neither was doing so with much relish. Osborn was from the world of Edith Wharton's *Age of Innocence*, where the sons of New York's social élite were expected to follow their fathers into the law and then the family business. It was never considered that they might strike out for themselves, especially into an area with such little chance of success and influence as the sciences. A trip to the Wild West to hunt fossils and battle Indians sounded like their last chance at adventure before settling into the rigid confines of adulthood. Osborn, however, suggested a warm-up trip to somewhere local like the Catskill Mountains that Guyot loved. The Catskill trip, based out of 'Wing-and-Wing' where Virginia could supply the explorers with food and snacks, went well enough that plans for the more ambitious Western excursion began.

Having drummed up support from the school, as well as the United States Army, the first Princeton Scientific Expedition got under way shortly after graduation in 1877. Dressing like cowboys and arming themselves to the teeth, the Yankee caballeros took off on their great lark West. With a contingent of eighteen students and two faculty members, the expedition visited Colorado, Utah and the Bridger Basin of Wyoming. The paleontological section – Osborn, Scott, Speir, R.H. Lynde and a student named Potter – collected fossils, watched out for Indians, and generally enjoyed themselves.[69] Though originally the trip was a way of spending a few last moments as unencumbered youths, Scott and Osborn found themselves being drawn to the notion that they might have a future as scientists. They worked hard to make their first scientific expedition an intellectually productive one. When it was over, Osborn published a pamphlet on some of the material he collected

Fig. 3 The Yankee caballeros – Henry Fairfield Osborn, Francis Spier and William 'Wick' Berryman Scott – dressed up for action out West in 1878.
Neg. No. 116113, courtesy Department of Library Services, American Museum of Natural History.

in Wyoming. The report shows that Guyot's synoptic technique as well as his mystical revelations about a fallen leaf had rubbed off. Osborn wrote that 'in these rocks are buried many generations of animals whose genealogy can often be clearly followed, and it is undeniable that here [the Bridger Basin] if anywhere, the records of an ancient land are written in characters which cannot be mistaken.'[70]

Graduate School

Following the Western expedition, Osborn took a year to examine the collection he had brought back with him. As part of his continuing effort to upgrade Princeton, McCosh encouraged Osborn and Scott to get their advanced degrees so they could join the staff. Their relationship with the president grew closer when they took his graduate classes. They were, in fact, the first Princeton graduate students – students McCosh referred to in his thick Scottish brogue as 'me bright young men.'[71] It had become tradition for Americans to go overseas to finish their studies at one or more of the renowned universities of the Old World. America had few opportunities for graduate study and so the journey was seen as a standard part of an education whether in the arts or sciences. Many American artists, such as James McNeill Whistler and Thomas Eakins, made the trip, as did a growing tide of students of natural history that included the Princeton boys. Scott was the first to go, initially to England, where he worked with Francis Maitland Balfour at Cambridge studying embryology, then with T.H. Huxley to work on comparative anatomy. After Britain he traveled to Heidelberg, Germany for more comparative anatomy with Carl Gegenbaur.[72]

Wick kept Harry apprised of life overseas, including information on the school, the various customs as well as the curious case of the German student. This individual committed suicide by cyanide poisoning after being arrested for having 'stolen something from one of the laboratories.' Scott shook his head in amazement. 'Who on earth,' he asked Osborn rhetorically, 'but a German would do such a thing?' The more Scott got to know Germans, 'the less I like them.'[73] When Scott returned home in 1880, McCosh made him his assistant. As for Osborn, his work on the Wyoming material complete, he left for New York in 1878 to study anatomy and physiology with William Welch at Bellevue Hospital. The next year, after some foot dragging by his father and pleading by himself, he too headed to Europe, following Scott's trail, first with Balfour and then Huxley.

In England, Henry Fairfield was on his own for the first time in his life. He found solace in hard work and letters home, especially to his brother William Church. The youngest Osborn boy had entered Princeton and the oldest gave advice on how to get through. As he had chaperoned William around Europe, Fairfield would guide him through college. Besides telling him how to behave, and the importance of study and hard work, Fairfield reminded William about the contract they had signed with their father in 1873. They had vowed never to begin smoking, else they would lose their inheritance. Henry added that smoking 'defiles' the body.[74] For himself, he was realizing how tough it was in Britain, but also how that toughness would result in a better education. English universities, he would later remember, were very good, but because of arcane rules and regulations it was difficult for Americans to work there. Overcoming these obstacles and pushing on ahead in the face of adversity was a major accomplishment which was rewarded with an excellent education.[75]

All was not hard work in England, however. Henry Fairfield did find time for some less intellectual pursuits, partaking of the quintessential aristocratic sport of fox hunting – though he fell off his horse and wrenched his back. He also received wisdom on what made Americans unique from 'the young Oxfordian, Oscar Wilde.'[76] Lodging at 17 Half Moon Street in London, Osborn quickly immersed himself in English society and became a regular on the party circuit. On an endless round of dinners and teas he flirted with pretty English girls, but did not think they 'compare with the Americans.' While he enjoyed spending time with them, he was not 'buccaneering' for a wealthy European wife, but a European education that would allow him to become a leader in American science. He also found comfort in the sermons on hellfire and damnation he was attending at the various church services in England (though his mother scolded him for often working on the Sabbath instead of going to church). During the sojourn in England, Henry Fairfield was beginning to get a sense of who he was, the special role life had cut out for him, and the importance of effort and struggle as a way to accomplishment. He wanted to hand this knowledge down to his brother so that he too could tread his own path to eternity. He told William that 'great men live ahead of their age – you can only get ahead of yours by rising above it.'[77]

Osborn enjoyed his time in England, particularly studying with Huxley. Darwin's bulldog had a facility for drawing explanatory pictures on the blackboard which made difficult points clearer, and engaged in good-natured banter with his students. Huxley once told a student that his anatomical drawing 'reminds me more than anything else of Cologne cathedral in the dark.'[78] Besides simple facts, Huxley taught his students that theory must always be based upon a wide variety of knowledge and experience. He told them they had to obtain 'a very broad view of the animal kingdom before [attempting] to enter the plane of higher generalization.'[79] While laboratory work was an integral part of his classes, Huxley – by then beginning to show signs of age – rarely appeared in the lab himself, allowing assistants to oversee the student's work.[80] He did, however, bring, unannounced, Charles Darwin to class for a sort of 'show-and-tell' event. The assembled class gaped as Huxley walked the father of modern evolutionary theory in. When Huxley introduced the legend to the young American, Osborn was speechless, barely able to do more than stare into the old man's eyes. When class ended Darwin shambled out, never to be seen again. Like a sighting of the Loch Ness monster it was all over before anyone had time to comprehend this sudden appearance. The moment stuck with Osborn the rest of his life. He told everyone about the encounter and lamented to his best friend Wick that 'I immediately thought how I wish you could have shared this pleasure with me.'[81]

Huxley himself was not quite as elusive; he and Osborn developed a relationship outside of class that included inviting Osborn to his 'Tall Teas.' Social gatherings at Huxley's home often included the glitterati of English arts, science and politics. At one dinner, Huxley drew him over to the fire to dispense some advice. As the logs crackled he told Osborn that it would be a good idea to go study with the powerful Yale paleontologist Othniel Charles Marsh upon his return to America, and to work to build America's scientific capabilities. Huxley was mightily impressed with America and its possibilities and wanted Osborn to play his part.[82] He referred to Osborn as 'pleasant … & well informed,' and thought of him as a disciple who would bring the 'Huxley Method' – the conversion of geology-based paleontology to biology-based paleontology – back to the colonies.[83] It was a role Osborn was happy to assume. In the end, however, as much as Osborn admired Huxley and adopted his reliance on empiricism and the importance of scientific education, on philosophical and theological issues they were two ships passing in the night. Huxley did not subscribe to the types of metaphysical beliefs that Osborn did.

After the year in England, Osborn was forced to put his scientific career on hold. William Henry Osborn was not overjoyed at his son's decision to become an intellectual instead of a businessman. The father was reluctant to allow Harry, now known as Henry, to go to on further Western expeditions or to England, thinking it not worth the effort. Henry asked Wick Scott to write his father (which he did several times) to explain how important it all was. Henry told Scott that a life of science was 'not a mere whim … but really a serious matter [to him].'[84] Initially William Henry told Scott to keep his nose out of places it did not belong. The elder Osborn had recently acquired the Chicago, Nashville and New Orleans Railroad, and wanted his son to help him run it. William hoped that a taste of big business would tempt him away from science. While in London Henry Fairfield had apologized to his father for not having a good business sense. He often had trouble keeping up his accounts and lost track of bills. But when asked, the young man dutifully obliged and entered the railroad business. After several months William Henry realized how much science meant to his son, and began to give in.

Released, Henry Fairfield lost no opportunity to naturalize. He hunted alligators in Louisiana and collected birds and birds' eggs in Texas. Drawing little alligators on his letter, he told his brother that the best place to shoot one of the mighty reptiles was in or above the eye.[85] Letters of support and encouragement from several reputable scientists in the United States and United Kingdom (engineered from behind the scenes by McCosh, who did not want all his hard work to go for naught) helped persuade William Henry to let go, though not completely. Resigned to the notion that his son was not going to study law or take over the family business, he built Henry a private library on the grounds at Garrisons and insisted that his son build an empire in the pursuit of knowledge. Years later, in a typical compliment, William Henry is said to have boasted of his son that 'maybe Fairfield would not have made a great business career, but he can already write a book you cannot read!'[86] Father Osborn would always be skeptical of his son's scientific career, but he tried to help him advance it any way he could. In 1880 New York's American Museum of Natural History offered Henry Fairfield a job as geologist, but he told museum director Albert Bickmore that he could not accept such a post while Edward Drinker Cope of Philadelphia and Othniel Charles Marsh of Yale held such overriding power in the field. If Fairfield was not to be a Captain of Industry he would at least be a Captain of Science, and tackling 'Scylla and Charybdis' (Marsh and Cope) at that point was a suicidal effort. Instead he held out for Princeton and the chance to build a power base there without interference.[87] William bankrolled his son's work, even paying his Princeton salary, when he was finally made assistant professor of natural science in 1881, because the school could not afford to do so.[88] With this position attained, he set out to make something of himself.

Loulu

As well as becoming a professional scientist in 1881, Henry married Lucretia 'Loulu' Thatcher Perry – a descendant of the Commodore Perry who opened Asia to the United States in the 1840s. She was not, however, the only lady in his life. During his stay in England Osborn had made trips to Germany in order to learn the language and inquire into possible places to study. There was more to hold him there than the language. In Coburg he lived with the Schindhelm family and through them met the Broderotti family and their fetching, busty daughter Clotilde. Henry and another American student named Slaughter spent a good deal of time with the Broderottis and he flirted heavily with her. He was attracted to her enough that he struggled through a special American-style dinner she prepared for them. Unfortunately, her cooking skills were lacking. Not wanting to offend her, he and Slaughter choked down the wretched meal, smiling the whole time. He could not 'refuse anything prepared by the hand of the lovely Clotilde.' That night Osborn mused to his diary, 'so much will men do for a fair hand!'[89] He thought her beautiful, and enjoyed the pursuit, but as he told Wick Scott, there was 'what I consider an insuperable barrier' to them going further: she was a Catholic.[90]

Anti-Catholic feeling in America, at its height from 1830 until 1860, saw a resurgence in the 1870s. The Catholic Church was seen as inimical to the American republic because of its perceived anti-democratic ways. Nativists saw Catholics as degenerates both because of their 'Romish' religion and because they were mostly Irish and Italian immigrants. As such they were a stain upon the Anglo-Saxon heritage and the moral fiber of the country. Lurid stories about the goings-on inside churches and convents made even nuns and priests suspect. To many Protestants, Catholics could never be good Americans.[91] The Osborn family was not immune to these feelings. During the Civil War William Henry Osborn had met a Sister of Charity and donated money to her hospital for wounded soldiers. But when he gave money to a school for renovations and improvements the contract stipulated that the 'said school house shall not be used for the holding of any Roman Catholic service whatsoever at any time.'[92] Thus for Henry Fairfield love may have bloomed for the sweet Fräulein, but her religion left him no choice, and he walked away.

Loulu Perry was a different story. The Osborn and Perry families had been acquaintances for years. Virginia Osborn and Loulu were, in fact, already friends and often dined together. Loulu was a first cousin to Virginia's niece Sarah MacWorter Sturges. Henry had known her for years, but their growing intimacy was not apparent to either family. When he told his mother that they had become engaged she was overjoyed, but registered shock at her ignorance that a relationship had been brewing all those years. Loulu was thus a known commodity, already well liked by the groom's mother, and there was no chance of an inappropriate religious proclivity raising its head. Virginia and Lucretia 'both wished to live the same life – a life with Christ in God,' and so the pairing was instantly accepted and blessed.[93]

The wedding was held at the Protestant Episcopal Church on Governor's Island, New York because Lucretia's father, an army general, was stationed there.[94] After their honeymoon, the Henry Fairfield Osborns returned to a gala party at the recently opened Castle Rock mansion on the Garrisons estate. While they waited for their home to be built in Princeton, they set up a homestead across the street from the university press, within walking distance of Henry's office and classrooms. Christening it 'La Casita', he had to endure its prompt renaming as 'La Casita on the backstreeta' by local campus wags.[95]

Henry and Loulu enjoyed a genuinely happy marriage full of affection and love. Not long after the wedding, and his settling down into the role of hard-working Princeton professor, he was wallowing in mounds of paperwork and research. None the less, he pushed it all aside to write her: 'here are a few lines of love to tell you of how you are in my thoughts and close in my heart this morning.'[96] When she left for Castle Rock later that spring, leaving him behind to work, he immediately went back to his office and wrote her of how watching the train take her away affected him. 'Here are a few lines of deepest love to tell you how bottomless everything seemed as the train pulled out of the depot and separated all that I have and love.'[97] In the second year of his marriage, Henry wrote his mother telling her how important she and his late sister had been in his life and how they had indirectly chosen his bride for him. 'Thank you for the appreciation you and sister have given me of true womanhood.'[98] His mother was the yardstick by which he measured all women. He could never have married a woman who was his mother's 'inferior.' Loulu, besides being the only woman he ever met who was on his mother's level, made up for the 'gap' left in his soul after his sister died.[99]

The task of coming to terms with his sister's death in 1874 was compounded by the death of his brother Frederick the same year. Frederick, along with his friend 'Teddy' Roosevelt, had been the budding naturalists. It had been Frederick, and not Fairfield, whose childhood dream it had been to become a scientist. His father had even brought Frederick together with some people at the American Museum of Natural History as a way of introducing him to the profession. Father Osborn had seen Freddie as the intellectual and Fairfie as the son who would carry on the railroad business. When the double tragedy hit, William reportedly lamented that 'we have lost the two children of genius.'[100] William Henry's lack of confidence in Henry Fairfield's intellectual promise did not make caring for him through his extended illness easy for the son. Though he still had twelve years to live, William Henry's chronic condition left him bed-ridden at regular intervals. Henry realized that his father was slowly dying and tried in vain to comfort him. 'His nerves are all shattered and beyond control,' he confided to Willie. He was also getting difficult to handle and would often yell at Henry deliriously as if he were someone else. It caused the young man great distress, but he loved his father dearly and sat with him, weathering the abuse.

Although he had trained himself not to break down every August (the anniversary of his siblings' deaths), the memory was a painful one that continued to haunt him. Having been away at college when the incidents occurred, Henry felt some responsibility.[101] To compensate, Henry took a close interest in William's career, and as head of the family took responsibility as his guide. He steered his brother away from the railroad and strongly suggested that William Church go into politics and put his law training to positive, creative use. He told his brother that there was no profession available for an honest man which provided such a 'useful and promising life.'[102]

In addition to holding the family together and seeing to William's future, Henry had a career of his own to worry about. In order to care for his father he turned down Wick Scott's invitation for a field trip out West. These responsibilities did not keep him completely idle, however. Between 1883 and 1886 Osborn and James McCosh collaborated on studies of the brain, vision and mental cognition. As a Princeton professor, Osborn began building a career as a scientific leader, making contacts in the burgeoning field of biology. In the workaholic tradition of his father, Henry taught embryology and psychology, wrote papers and gave demonstrations. He worked at both assisting McCosh's program of making science a major curricular focus at the school, and building up the intellectual life at Princeton by sponsoring student meetings and discussions.[103] He also immersed himself in the nonacademic life of the school. As chairman of the Graduate Committee on Athletics, for example, he rebuilt the Princeton athletics system and was a tireless booster.[104] He also began several clubs, including a Sketch Club where students and faculty members could practice drawing the human form from life. Osborn had a natural talent for drawing and the club was a way of exercising it.[105]

Edward Drinker Cope

Henry's scientific interests were also maturing and he began to look to new horizons. Studying with Balfour and Huxley had planted seeds in his mind about how he wanted to proceed with his career. Huxley had been looking at evolutionary relationships, which interested Osborn. Osborn was increasingly drawn to studying development as a way of determining homologies and phylogenetic relationships in evolution. Morphology (the study of an organism's structure) comprised a number of different disciplines: comparative anatomy, cytology, embryology and paleontology. In the years following the publication of Darwin's *Origin of Species* (1859) morphology became the primary inductive methodology to search for large-scale answers to questions about evolution. As the search for a unified theory of evolution intensified, morphology became the basic biological approach to studying living things. The aims of morphology were to discover evidence for the unity, or relationship, of diverse forms. Also, morphology sought to reconstruct family trees, or phylogenies (the course of an entire group, or phyla, over time). This preoccupation with outward form and group histories led to the belief that the development of an individual organism, particularly in its embryonic stage, was a history of how the entire group an individual belonged to had evolved.[106] This was the methodology Osborn used. However, after publishing a few papers on the subject in 1883, he discovered that he lacked the dexterity needed for work as an embryologist or cytologist, which required the manipulation of microscopic embryos and individual cells. Instead he turned his attention to the fossils he had become comfortable with from the Western expeditions.[107] By doing so, in 1885 he found his calling: he would spend the rest of career attempting to explain how the fossil record showed the clear evidence of the development and evolution of species and how organisms advanced.

To get his career as a paleontologist off to a good start, Henry Fairfield Osborn went back overseas to study paleontology in Germany. His serious work on evolution began in the late 1880s, a period in which he collaborated with William Berryman Scott on a paper on fossil mammals and their relationship to evolutionary questions. More crucially, in 1885 he resumed a relationship with Edward Drinker Cope (1840–97), one of America's fore-

Fig. 4 Edward Drinker Cope – the mad genius of Philadelphia – was Osborn's mentor and friend. His war with O.C. Marsh helped fill the coffers of American museums with an unprecedented collection of fossils.

Neg. No. 116520, courtesy Department of Library Services, American Museum of Natural History.

most paleontologists.[108] Cope was born into a wealthy Philadelphia, Quaker family; his father was an amateur horticulturist who interested his son in nature. In 1860 he entered the University of Pennsylvania as a student of the pioneering Philadelphia naturalist and polymath Joseph Leidy (1823–91). In 1863 Cope went to Europe to study for a brief period, after which he accepted the post of professor of natural science at Haverford College, Pennsylvania. Cope's true vocation was not teaching, but research, so in 1867 he left Haverford to follow that path. His family's wealth allowed him to concentrate on pure research with gusto. Drawn to the study of fossil vertebrate life and its development, his first article on evolution appeared in 1868. Cope was a prodigious worker, publishing thirty-one papers by the time he was twenty-one years old. He spent most of the years between 1871 and 1879 working the fossil fields of the West on the various geological survey missions then active in the area. Cope amassed enough material to produce a major treatise on the Western fossil fauna as *The Vertebrata of the Cretaceous Formations of the West* (1875) and later *The Vertebrata of the Tertiary Formations of the West* (1884). This ground-breaking work was the first comprehensive study of extinct animal life in America, and helped push dates for the antiquity of mammal life back further than they had ever been.

Though Philadelphians, Cope and Leidy were opposites in almost every other respect. One of the founders of paleontology in America, Leidy had a hyperactive mind that approached biology in a scattershot fashion. He would become interested in a field, do intense, rigorous and often pioneering research and then bounce to the next topic which caught his attention. He would then leave the previous field to be exploited by others. He did ground-breaking work in paleontology, parasitology, microscopy, anatomy and public health. He described the first dinosaur discovered in America, the Hadrosaur of Haddonfield, New Jersey, and was the first to put forward the notion that the fossils of the United States represented, not lone monsters and oddities, but entire faunas that covered the continent in the same way modern animals do. Whereas Cope was brash, outspoken, combative and reckless in his research – a sort of mad genius – Leidy was modest to the point of obsequiousness, kept his views to himself, and worked meticulously to build up mountains of facts. Leidy shied away from theorizing, but instead was a great descriptive scientist, carefully cataloging morphological minutiae.[109]

In 1872 Cope was asked by F.V. Hayden to join a Western survey mission he was running. Hayden had taken control of the recently created United States Geological and Geographical Survey, the organization that eventually became the United States Geological Survey (USGS). The survey was making a scientific inventory of the Western states for use by the government in Washington. Cope eventually made his way to Wyoming and the Bridger Basin.[110] Due to its geological condition the New World was a treasure trove of fossil material. The Native Americans had long since known of fossils and used them as decorations and religious ritual objects.[111] When Europeans arrived they also became fascinated with the ancient relics. Collecting became a pastime of natural historians and intellectual dilettantes as early as the 1700s. Bones found along the Hudson River and throughout New England in the colonial period were thought to be the remains of biblical giants killed off during Noah's flood. By the eighteenth century the bones were being seen for what they were, the remains of long-dead animals that had once roamed the American landscape. Philadelphia, a center of intellectual activity in the colonial era, became a center of fossil research, with naturalist Casper Wistar (1761–1818) doing some of the first technical studies of the stony curiosities.[112]

Edward Drinker Cope's appetite for fossils to study led him into a collecting 'bone war' with his rival from Yale College, Othniel Charles Marsh. The two engaged in a twenty-year battle to see who could collect, name and describe the most spectacular fossils.[113] Like Cope, Marsh was independently wealthy and used his fortune to finance his scientific pursuits. Not an evolutionist originally, Marsh sat in on Arnold Guyot's early lectures in America, studied at Yale and then went to England in 1862 to work with Charles Lyell and T.H. Huxley. There he became interested in studying descent, after which he accepted Darwinian evolution as a scientific fact. Upon his return he took the position, the first in America, of professor of paleontology at Yale in 1865.[114] Marsh began building a formidable fossil collection to use as a foundation from which to study questions of evolution. He began building his reputation in 1872 by making the bird–reptile–dinosaur link through his discovery of fossil birds with reptilian characteristics. His career solidified with the pioneering work done on fossil horses which so impressed T.H. Huxley.[115] He and Cope had become the titans of American paleontology that Henry Osborn feared enough at the time to turn down a position at the American Museum of Natural History.

Marsh and Cope each had collecting teams in the West feverishly digging and shipping back fossils to New Haven and Philadelphia. The teams fought over dig sites, bribed railroad agents and shippers and generally worked to undermine each other's efforts. Building on an animosity going back to the beginning of their relationship, a genuine hatred developed between the two men. In addition to the fossil competition, a sticking point between the two was the geological surveys. Cope had been brought into the government fold – which meant government financial support – by F.V. Hayden, while Marsh's entrée was through John Wesley Powell. A struggle broke out over who would become director of the new United States Geological Survey and gain its considerable political power. The two candidates, Hayden and Powell, soon had factions forming around them. Both Cope and Arnold Guyot backed Hayden for the job, while Marsh went with Powell.

John Wesley Powell (1834–1902) was basking in the limelight of his romantic, adventurous exploration of the Grand Canyon which had caught the country's imagination. He eventually was handed the position as head of the Geological Survey in 1881, the same year Osborn received his professorship at Princeton. Becoming head of the Smithsonian Institution's Bureau of Ethnology as well, Powell postulated that since some Indian artifacts seemed to be linked to the movements of glaciers, and thus the Ice Age (a concept worked out by Swiss geologists like Arnold Guyot and Louis Agassiz), studying the geology of this phenomenon would reveal facts about the origins of mankind and how they arrived in the New World. For Powell, such a study had moral as well as intellectual implications.[116]

With his benefactor gone, Cope's power base was weakened, and Marsh, now in the political ascendancy at the USGS, saw to it that Cope's federal funding was cut. This infuriated Cope and added insult to injury, for he had recently lost most of his inheritance to a series of disastrous investments. This drastic change of fortune limited his ability to collect or buy material.[117] Because of this frenzied rivalry, however, the fossil record was being filled in rapidly. Where Europeans had only scant pages from the book of life, the Americans now had volumes, and Osborn would make this book the cornerstone of his work.

It was towards the end of this period that Osborn met Cope and was swept up in the conflict. Arnold Guyot enjoyed a good rapport with the volatile fossil hunter, as well as with Cope's one-time mentor Joseph Leidy, and had introduced him to Osborn and Scott during the Western expeditions as up-and-coming naturalists.[118] Osborn asked Cope for advice in dealing with fossil dealers and professional collectors, a topic Cope had much experience with.

With an introduction from Guyot, Osborn and Scott stopped in the 'City of Brotherly Love' in 1877 on their way out West. The standoffish Cope at first regarded them with suspicion, but the two Princeton boys eventually won him over, and they became friends. Though he acknowledged his mentor's quirks and idiosyncrasies (Cope, though a Quaker, was combative, defensive and possessed a self-confidence which could easily look like arrogance), Osborn admired Cope and distrusted Marsh. When Osborn and Scott visited Yale in the late 1870s to consult the specimens, Marsh had all the most important material locked away and restricted them to certain parts of the museum. In order to keep an eye on them, Marsh silently followed the two from a distance in a pair of carpet slippers as they wandered the Peabody Museum's halls.[119] Marsh also had a reputation for not giving his assistants their due. It was claimed that he often took credit for work done by his assistants and students and resisted letting them develop their own ideas, instead setting them to do research projects which enhanced his career. Osborn was aware of the allegations. The semester before he began at Princeton as assistant professor, he spent time at Johns Hopkins University's Chesapeake Zoological Station doing research. There he heard the horror stories about

Marsh's behavior as well as similar legends about Louis Agassiz of Harvard. The Zoological Station's director, William K. Brooks, delighted in regaling the wide-eyed Osborn with personal stories about how much of a bore the legendary Agassiz was.[120]

Cope wanted the youngsters Osborn and Scott to help him undermine Marsh's position, and in the late 1880s and early 1890s worked to enlist their help. Cope feared that Marsh was attempting a coup to take control of the Geological Survey, the National Academy of Sciences, and American paleontology in general. The war even spread to the media. In 1890 a series of newspaper articles ran in the *New York Herald* that documented the conflict and listed the grievances of both parties. The articles appeared originally at the prompting of Cope and a ne'er-do-well 'journalist,' William Hoisea Ballou. The public airing of paleontological dirty laundry dragged both Osborn and Wick Scott into the fray against their wills, and soiled all their faces.[121] In addition, an election for the librarian of the American Philosophical Society came up and Cope feared the candidate, L.S. Lyman, was too close to Marsh and would keep Cope's papers out of the Society's publications. He wanted Osborn and Scott to vote against Lyman and rally support to oust Marsh from his position of power. Cope called Marsh an 'imposter and tyrant' who had 'given into coercion and intimidation.'[122] To complicate matters, Osborn respected Leidy, who liked Marsh and disapproved of Cope. Leidy had given Osborn an entrée into various loci of power by bringing him closer to Cope, and introducing him to Marsh as well as F.V. Hayden and the work going on out West. Osborn hated Marsh and told Wick Scott flatly, 'I shall never serve under Marsh even if a fine position is offered me.'[123] American paleontology was an entangled soap opera of competing egos, power plays and animosities which Osborn, as the newcomer, was forced to tread like a soldier negotiating a minefield. He played it carefully by supporting Cope, but not antagonizing Marsh, at least not until he had some professional security and clout of his own. It was a tribute to Osborn's growing political sense that he did it so adroitly. Osborn remained faithfully in Cope's camp his entire life and the Philadelphian's evolution theory became a central part of Osborn's own early thinking on evolution, as would Cope's views on society, race relations and salvation. These aspects of their relationship will be discussed later.

Conclusion

By the end of the 1800s Henry Fairfield Osborn was beginning to consider questions of evolution as the core of his scientific research. He had been exposed to the teachings of several strong presences in his life which had a direct effect upon his world-view. His father taught him self-possession, the importance of social respectability, the need to be a leader in his field. His mother imparted the need to create his own individual relationship with the Almighty and the importance of the struggle not only for material success, but for spiritual salvation as well. At Princeton, Arnold Guyot and James McCosh showed him that he could be simultaneously a good Christian and a good scientist. McCosh opened him to the idea that a built-in human intuition existed which affected behavior and ability, emphasizing the idea of a mental, as well as physical, evolution. Time spent with Guyot and McCosh, as well as Balfour and Huxley, had made him an empiricist preoccupied with the need to have factual support for theory. For Osborn the fossil explosion of the American West would supply the record which could be read, as Guyot said it could, to reveal all the truths of development, both physically and spiritually, of life on earth. By the end of the century, notions learned as a child and a young man had become the basis and motivation for his study of the origins of man in order to learn not only about the past, but also about himself.

Notes

1. *American Dictionary of Biography* (New York: Oxford University Press, 1999), Vol. 16, 791.
2. Virginia Sturges to her parents, 3/20/1853. Also see other correspondence in the Virginia Sturges Osborn Papers, folders I & II, J.P. Morgan Library, New York. Hereafter VSO Morg.
3. Virginia Sturges to her mother, 4/24/1853, VSO Morg.
4. Virginia Osborn to her sister, 3/6/1856, folder II, VSO Morg.
5. *History of the Illinois Central Railroad Company* (Chicago: Railroad Historical Company, 1900), 68.
6. *The Pedigree of Henry Fairfield Osborn* (New York: privately printed, 1919), 2.
7. For Osborn's career with the Illinois Central see ibid. Also see John Stover, *Iron Road to the West* (New York: Columbia University Press, 1978), and Paul Wallace Gates, *The Illinois Central Railroad and its Colonization Work* (Cambridge: Harvard University Press, 1934).
8. Herbert J. Satterlee, *J.P. Morgan: an intimate portrait* (New York: Macmillan Co., 1940), 102.
9. Jean Strauss, *Morgan: American Financier* (New York: Random House, 1999), 81.
10. See Genealogy folder, Osborn Papers, New York Historical Society. Hereafter OSNY.
11. Virginia to William Osborn, 8/7/1859, folder III, VSO Morg.
12. Virginia Osborn to William Henry Osborn, 9/7/1859, folder III, VSO Morg.
13. *Time Magazine* (12/31/1928).
14. Henry Fairfield Osborn to Henry Sturges [8/2/1867?], folder VIII, VSO Morg.
15. HFO, *Fairfield Reminiscences* (n.d.), possibly early 1930s.
16. Virginia Osborn to Lucy Sturges, 5/6/1861, box I, OSNY.
17. Strauss, *Morgan*, 92.

18. HFO, *Fairfield Reminiscences of Eighteen Sixty Seven* (New York: privately printed, 1935).
19. *Lincoln in the Telegraph Office* (1907).
20. Virginia Osborn to William H. Osborn, 5/23/1871, folder V, VSO Morg.
21. William K. Gregory, *Biographical Memoir of Henry Fairfield Osborn* (Washington, DC: National Academy of Sciences, 1938), 54.
22. G.W. Stone to HFO, 5/3/1871, and HFO to W.H. Osborn, 9/12/1871, box I, OSNY.
23. Virginia Osborn to her mother, 6/24/1871, VSO Morg.
24. Notebook of Virginia Osborn, 1872, box XXIV, OSNY.
25. Notebook of Virginia Sturges, 1844, box XXIV, OSNY.
26. Notebook of Virginia Sturges, 1851, box XXIV, OSNY.
27. Virginia Osborn to Jonathan Sturges, 5/20/1857, box I, OSNY.
28. HFO, *Fairfield Reminiscences*.
29. Virginia Osborn to Jonathan Sturges, 5/20/1857, box I, OSNY.
30. Craig James Hazen, *The Village Enlightenment in America* (Chicago: University of Chicago Press, 2000), 6.
31. Quoted in *Christian Work and Evangelist Magazine* (5/27/1905).
32. Elizabeth C. Hobson, *Recollections of a Happy Life* (New York: G.P. Putnam's Sons, 1916).
33. The undated, handwritten manuscript is in folder X, VSO Morg., along with her other religious writings.
34. Barbara Welter, 'The Feminization of American Religion, 1800–1860,' in Mary Hartman and Lois Banner, eds, *Clio's Consciousness Raised* (New York: Harper Torch Books, 1973).
35. Virginia Osborn to Amelia Sturges [1860?], box 5, folder 3, Osborn/Dodge Papers, Manuscripts Division, Department of Rare Books and Special Collections, Princeton University Library. Hereafter PU.
36. Virginia Osborn to Fanny Morgan, 8/25/1885, box 5, folder 5, Osborn/Dodge Papers, PU.
37. Virginia Sturges Osborn to Henry Sturges, 1/1/1872, folder V, VSO Morg.
38. Gregory, '*Biographical Memoir*,' 55.
39. Helen Warren and Florence Milligan, *Fifty Years of Princeton '77* (Princeton: Princeton University Press, 1928), 222.
40. William Berryman Scott, *Some Memories of a Paleontologist* (Princeton: Princeton University Press, 1939), 36.
41. Box XX, Princeton University folder, OSNY.
42. HFO, personal scrapbook, AC#26, Princeton University Archives, Seeley G. Mudd Manuscript Library, Princeton University. Hereafter MUDD.
43. *New York Times* (11/20/1932).
44. See family correspondence in box I, OSNY.
45. *Princeton Alumni Weekly*, 30:29 (1930).
46. Virginia Osborn to HFO, 2/5/1874, box I, OSNY.
47. Virginia Osborn to HFO, 4/25/1875, box II, OSNY.
48. Virginia Osborn to HFO, 11/8/1874, box I, OSNY.
49. Virginia Osborn to HFO, 2/1/1874, box I, OSNY.
50. Virginia Osborn to HFO, 12/19/1875, box II, OSNY.
51. Virginia Osborn to HFO, 12/2/1873, box I, OSNY.
52. Virginia Osborn to HFO, 10/31/1875, box II, OSNY.
53. Ibid.
54. HFO, *Creative Education* (New York: Charles Scribner's Sons, 1927), 24.

55. *Dictionary of Scientific Biography*, 599.
56. Faculty File, Arnold Guyot, box 208, folder I, MUDD.
57. 'Memoir of Arnold Guyot,' *Third Annual Report of the E.M. Museum of Geology and Archaeology* (June 1884).
58. Student Class Notes, 1859/60, Arnold Guyot Lecture Notes Collection, AC#52, box 17, folders II & III, MUDD.
59. Faculty File, Arnold Guyot, box 208, folder I, MUDD.
60. J. David Hoeveler, *James McCosh and the Scottish Intellectual Tradition* (Princeton: Princeton University Press, 1981).
61. *Princeton Sesquicentennial Celebration 1746–1896* (New York: Charles Scribner's Sons, 1898), 424–425.
62. George Schmidt, *Princeton and Rutgers: Two Colonial Colleges of New Jersey* (Princeton: Van Nostrand Co., Inc., 1964), 77–79.
63. William Milligan Sloane, ed., *The Life of James McCosh: a Record Chiefly Autobiographical* (New York: Charles Scribner's Sons, 1896), 166–175.
64. James R. Moore, *The Post-Darwinian Controversies* (Cambridge: Cambridge University Press, 1979), 246.
65. James McCosh, *The Religious Aspects of Evolution* (New York: Charles Scribner's Sons, 1890), ix.
66. Notebook of John Melick Van Dyke, 1870, Lecture Notes Collection, James McCosh, AC#52, box 32, MUDD.
67. Quoted in Gregory, '*Biographical Memoir*,' 56.
68. For this scene see Scott, *Some Memories*, 48 and HFO, *Creative Education*, 24–25.
69. Princeton University Scientific Expeditions Collection, AR#012, box 1, Geological expeditions, 1877–1886; and H.F. Osborn, W.B. Scott and F. Speir, *Paleontological Report of the Princeton Scientific Expedition of 1877*, vol. 1 (New York: S.W. Green, 1878).
70. HFO, *A Memoir Upon Loxolophodon and Uintatherium* (Princeton: 1881).
71. Quoted in Hoeveler, *James McCosh*, 285.
72. See Scott, *Some Memories*.
73. William Berryman Scott to HFO, 2/8/1879, Osborn Papers, William L. Clements Library, University of Michigan.
74. HFO to William Church Osborn, 9/10/1878, box 5, folder 11, Osborn/Dodge Papers, PU. See contract 12/24/1873, box I, folder 1873, OSNY.
75. HFO, *Impressions of the Great Naturalists* (New York: Charles Scribner's Sons, 1924), 73.
76. HFO et al., 'The American Now in the Making,' *The New York Times Magazine*, 5:1, 14 (1/17/1932).
77. HFO to William Church Osborn, 9/10/1878, box 5, folder 11, Osborn/Dodge Papers, PU.
78. Diary of HFO, September, 1879, box XXIV, OSNY.
79. Quoted in HFO, *Impressions*, 82. And HFO to William Henry Osborn, 12/11/1879, box II, OSNY.
80. Leonard Huxley, ed., *The Life and Letters of Thomas H. Huxley* (New York: D. Appleton & Co., 1901), 435.
81. HFO to William Berryman Scott, 12/11/1879, box II, OSNY.
82. Diary of HFO February 8, 1880, box XXIV, OSNY.
83. Quoted in Adrian Desmond, *Huxley: From Devil's Disciple to Evolution's High Priest* (New York: Addison-Wesley, 1995), 510.
84. HFO to William Berryman Scott, 2/5/1879, box II, OSNY.

85. HFO to William Church Osborn, 4/1881, box 5, folder 11, Osborn/Dodge Papers, PU.
86. From the reminiscences of L.V.F. Randolph, March 1916, box XVII, OSNY.
87. HFO to William Berryman Scott, 7/11/1880, box II, OSNY. Scylla and Charybdis are a rock and whirlpool from Greek mythology that devour everything that comes near them.
88. Ronald Rainger, *An Agenda for Antiquity* (Tuscaloosa: University of Alabama, 1991), 34.
89. Diary of HFO, September 1879, box XXIV, OSNY.
90. HFO to William Berryman Scott, 9/13/1879, box II, OSNY.
91. Ray Allen Billington, *The Protestant Crusade* (New York: Macmillan Co., 1938).
92. For the nuns see Virginia Osborn to her mother, 12/16/1861, box I, OSNY. For the contract see 11/9/1867, box I, OSNY.
93. Virginia Osborn to Lucretia Perry, 4/9/1881, box III, OSNY.
94. *The Pedigree of Henry Fairfield Osborn*, 3.
95. Scott, *Some Memories*, 145–146.
96. HFO to Lucretia Osborn [?/?/1882], box III, OSNY.
97. HFO to Lucretia Osborn, 4/7/1882, box III, OSNY.
98. HFO to Virginia Osborn, 6/4/1882, box III, OSNY.
99. Ibid.
100. From the reminiscences of L.V.F. Randolph, March 1916, box XVII, OSNY.
101. HFO to W.C. Osborn, 7/2/1882, box 5, folder 11, Osborn/Dodge Papers, PU.
102. HFO to W.C. Osborn, 7/3/1882, box 5, folder 11, Osborn/Dodge Papers, PU.
103. Rainger, *An Agenda*, 25–33.
104. *Fifty Years at Princeton '77*, 223.
105. Scott, *Some Memories*, 147.
106. Garland Allen, *Life Sciences in the Twentieth Century* (New York: John Wiley and Sons, Inc., 1975), 1–7.
107. HFO, *Fifty-Two Years of Research* (New York: Scribners, 1930), 65.
108. Rainger, *An Agenda*, 37.
109. Leonard Warren, *Joseph Leidy: The Last Man Who Knew Everything* (New Haven: Yale University Press, 1998).
110. HFO, *Biographical Memoir of Edward Drinker Cope* (Washington, DC: National Academy of Sciences, 1929).
111. E.M. Kindle, 'American Indian Discoveries of Vertebrate Fossils,' *Journal of Paleontology*, 9 (1935), 449–452.
112. G.G. Simpson, 'The Beginnings of Vertebrate Paleontology in North America,' *Proceedings of the American Philosophical Society*, 86 (1943), 130–188. (Benjamin Franklin discounted extinction and thought the elephant bones being exhumed were those of some monstrous antediluvian carnivore.)
113. There are several works that cover the bone wars. Two of the better ones are Url Lanham, *The Bone Hunters* (New York: Columbia University Press, 1973), and David Wallace, *The Bone Hunter's Revenge* (New York: Houghton Mifflin Co., 1999).
114. John Ostrom and John McIntosh, *Marsh's Dinosaurs* (New Haven: Yale University Press, 1966), v.
115. Charles Schuchert and Clara Maclevene, *O.C. Marsh: Pioneer in Paleontology* (New Haven: Yale University Press, 1940).
116. Thomas G. Manning, *Government in Science: The United States Geological Survey* (University of Kentucky Press, 1967), 56–79.

117. Schuchert and Maclevene, *O.C. Marsh*, 252–261.
118. Manning, *Government in Science*, 54–55.
119. Edwin Colbert, *The Great Dinosaur Hunters and Their Discoveries* (New York: Dover Publications, 1984), 148.
120. HFO to Virginia Osborn, 9/26/1880, box 5, folder 10, Osborn/Dodge Papers, PU.
121. Wallace, *Bone Hunter's Revenge*, 209–254.
122. See letters from Cope to Scott, 1886–1890, box 12, folder 24, William Berryman Scott Papers, PU.
123. HFO to William Berryman Scott, 12/8/1878, box II, OSNY.

CHAPTER THREE

The Romantic Empiricist

The United States went through a period of self-identification in the nineteenth century. Continuing a program begun with the Revolution, intellectuals worked in earnest at throwing off the influence of old European standards. Americans were the New Adam in the Eden of the New World, taking on the challenges of the future without the dusty baggage of Old World sensibilities.[1] Walt Whitman wrote, 'you shall no longer take things at second or third hand … nor look through the eyes of the dead … you shall listen to all sides and filter them for yourself.'[2]

Along with other aspects of life, religion was being filtered anew as well. Americans in the mid-nineteenth century were still profoundly religious, but many were growing disillusioned with traditional forms of Christian worship. People stayed away from church in droves yet were increasingly interested in matters of the spirit. Historian Anne Rose argues that the heart of American Victorian culture was not the mechanical progress normally associated with it, but Romanticism. This is what powered the American impulse to search for secular approaches to questions once settled by traditional religious sources.[3] A host of alternative philosophies, from transcendentalism to spiritualism, Mormonism and Christian Science, as well as self-help movements and medical cures, sprang up, providing individuals a rich buffet of concepts with which to sate their spiritual appetites.

One of the more popular avenues in the pursuit of spirituality was the glorification of nature. Philosophers and writers from James Fenimore Cooper to Ralph Waldo Emerson, Henry David Thoreau and Herman Melville saw in the American wilderness a source of spiritual renewal and truth, not only about the nature of the universe but of their own natures. Artists like Winslow Homer, Asher B. Durand, and the Osborns' friend Frederick Edwin Church were captivated by the uniquely American landscape and its perceived recuperative powers. Church saw God's handiwork in nature and created landscapes which expressed its transcendent power. John James Audubon posed his avian subjects in a way that replicated human morality by posing his models like lovers. George Caleb Bingham portrayed frontier legend Daniel Boone as Moses leading the chosen American people on an exodus away from the bondage of Europe into the freedom and spiritual salvation of the wilderness of the West. The 'naturalism' of William Sydney Mount and the stark realism of Thomas Eakins were a part of the idea that the American experience could teach us much if we only looked closely enough.[4]

The philosophers and artists were not alone. Naturalists and scientists too were drawn to the woods for many of the same reasons Thoreau was. In addition to wanting to prove to their European cousins that they could do original thought-provoking work and advance the cause of science, there was something else, something more personal. The naturalists and fossil hunters who fanned out over the landscape sought not only to discover scientific facts about this alien world, but to find answers to the higher questions of life. Science was often seen as being in the service of religion. The early American geologist Edward Hitchcock, in his pioneering studies of the rocks and fossils of New England, was attempting to reconcile geology with scripture. Edward Drinker Cope's nemesis John Wesley Powell was driven to explore the Grand Canyon out of a desire to find personal redemption. The idea that nature had a distinct spirit of its own and was thus deserving of ethical consideration was the conclusion of naturalist William Bartram, son of America's first botanist, John Bartram. The younger Bartram's spiritual view of nature influenced writers like Coleridge and Cooper as well as philosophers Emerson and Thoreau.[5]

Thus the avenue ran both ways – science could serve art and philosophy as well. Many nineteenth-century religious movements made clear efforts to support their beliefs with science. Spiritualism, Christian Science and theosophy all did and saw no contradiction, in fact saw science and religion as compatible. Their work either incorporated current science or claimed that the scientific method of verifiable demonstration could prove the validity of their beliefs. As a metaphor for personal salvation the natural world could provide not only factual knowledge, which would enhance America's standing in the scientific world, but would cleanse people's souls as well. Though coming of age in the later part of the era, Henry Fairfield Osborn felt the same impulse to reach for metaphysical answers along non-religious avenues. In his case the pursuit of nature manifested itself in his study of paleontology and evolution.

Evolution Comes to America

There was little discussion of evolution (or transmutation as it was called) in America before the publication of Charles Darwin's *Origin of Species* in 1859, and there is debate about just how American intellectuals reacted to this new idea once it did arrive. Most were converted to a general belief in evolution, though not always to Darwin's central discovery – that natural selection was the driving force behind evolution. This dissatisfaction with natural selection caused some Americans to adopt the more comfortable explanations for organic change put forward earlier in the century by Frenchman Jean Baptiste Lamarck. He theorized that organisms change through the accumulation of hereditary traits brought on by adaptation to environmental pressure, and that changes to any aspect of the organism's structure were passed on to its offspring, creating a new species. For some, Lamarck's law of acquired characters was a simple, straightforward and easily understood alternative to Darwin's twin threat of selection and variation. Those less willing to dismiss Darwin's theory found support for it when in the 1880s German cytologist August Weismann proposed his 'germ plasm' theory of heredity. Weismann argued that only changes to the cells of the reproductive system of an organism – the germ plasm – could be passed on to offspring and that changes to the body, or somatic cells, could not. Darwinists argued that an accumulation of modified sex cells was the raw material upon which natural selection worked, in conjunction with other elements such as population pressures, and the natural variation of hereditary material. The 'neo-Darwinian' concept, which accepted natural selection as the central mechanism behind evolution, became the antidote for the slide back to Lamarckian mechanisms.

Some historians, such as Peter Bowler, have argued that lines between the neo-Lamarckians and neo-Darwinians were clearly drawn and amounted to a rivalry between opposing forces.[6] More recently, Ronald Numbers has argued that there is little evidence to support this conflict hypothesis. One problem, as Numbers points out, is that there was little consensus during the nineteenth century about what was meant by Darwinism (or Lamarckism for that matter). Many naturalists and other intellectuals claimed to be Darwinists, but what they actually believed about evolution varied, not only from each other, but from Darwin as well. As such, the term 'Darwinist' became synonymous with any evolutionist.[7]

While there may or may not have been a feud between neo-Lamarckians and neo-Darwinians, Henry Fairfield Osborn came down, at least initially, on the side of the Lamarckians. Though he was, and continued to be throughout his entire life, a great admirer of Darwin, his personal experiences, his belief in straight-line progress as well as his relationship with Edward Drinker Cope drew him away from strict Darwinism. Where Osborn's thinking remained Darwinian (as it did for others) was in his belief in the importance of struggle. It would be necessary to rearrange the scientific evidence to answer both scientific and metaphysical questions. In this way he would blend German Romanticism with British empiricism in a pragmatic American fashion to get what he wanted. He became a Romantic empiricist, not unlike Ernst Haeckel or even Madame Blavatsky.

It was not just biologists and theologians who saw the impact of evolution on domains outside of science. Other intellectuals saw evolution as having wider implications. Some applied it to society, some to economics and politics, some to human equality. Some – like Charles Sanders Peirce, William James and members of the Massachusetts-based 'Metaphysical Club' – saw evolution as connected to religion and philosophy. As Osborn himself later would, these metaphysicians strove to reconcile evolution with religion, and empiricism with idealism. They shared with Osborn a belief in the fact of evolution and the supremacy of science in solving problems over religious dogma. They were not totally in agreement, however. Peirce and James – the founders, along with John Dewey, of pragmatism – were far more comfortable with the role of chance and fortuitous elements than Osborn ever would be. Where Osborn would accept a certain amount of determinism or predestination, they rejected it. Peirce saw chance leading to order, while Osborn saw it leading to chaos. Peirce did share with Osborn misgivings that natural selection was not the answer to how evolution actually worked, and a belief in the importance of verifiable evidence. James saw a place for personal effort and will-power, and wrestled with the knotty problem of determinism versus free will as Osborn did. Both were uncomfortable with a universe where human freedom counted for little. In the end, James claimed that having both determinism and free will was nonsense.[8]

Trying to work out Darwinism in the human social world as well as the natural world was a common intellectual pursuit. The social Darwinist William Graham Sumner believed that Darwin taught that human society followed the laws of struggling to survive that existed in nature, and that there was no such thing as human equality or human rights. On the other hand, the pioneering sociologist Lester Frank Ward countered that intelligence and ability were not conditions of rank or good breeding.[9] Osborn imbibed much of this discussion, but had a more powerful influence.

Cope and Evolution

Even more than Arnold Guyot, James McCosh or any other intellectual in the formation of Osborn's theory of evolution, Edward Drinker Cope was pivotal. Cope began developing his evolutionary thinking in the mid-1860s at about the same time as Alpheus Hyatt, a promising student of Harvard's Louis Agassiz, was doing likewise. Hyatt held a belief, common at the time, in embryonic development where the embryo passed through stages similar to those of the adult form. Cope did so as well, and so the two formed an alliance.

Cope began expounding his evolutionary ideas in *On the Origin of Genera* (1868) and *The Method of Creation* (1871), as well as hundreds of articles and other works. He teamed up with Hyatt and Alpheus Packard Jr of Brown University, to help edit the journal *American Naturalist*, which became the house organ for their growing movement. They saw Darwin's concept of natural selection as troubling not only because it seemed difficult to explain and overly complex, but because of what they perceived as the theory's inherent atheism and chaotic nature. While they believed that natural selection was a part of the evolutionary process, they did not see it, as the Darwinians did, as the primary agent of organic change. The alternative they put together, dubbed neo-Lamarckism by Hyatt in 1884, contained the basic Lamarckian law of acquired characters enriched with a few of their own ideas.[10] These included the 'law' of acceleration and retardation which was linked to the use and disuse of parts and their effect upon the embryo. Organisms, through their interaction with the environment, would acquire adaptations (the law of acceleration) or lose those already acquired but no longer used (the law of retardation). For Cope an organism's struggle with the elements was central to its evolution, as was a mystical internal growth force he called Bathmism. He saw 'intelligent direction' pushing adaptations to mesh with the environment.[11]

For Cope (and other neo-Lamarckians to varying degrees) there were two basic forces at work in evolution. The first was the organism's interaction with its environment. The second, which at the embryonic level he called Mnemogenesis, was the organism's consciousness. There existed a sort of super-consciousness in the universe which exerted pressure on the individual's consciousness in a process he called Archaesthetism. Cope told his father, 'I can prove, I think, the pre-existence of mind, i.e., as existing in living matter before it has developed complicated structure.' He went on: 'it is a satisfaction to me to be able to prove the fatherhood of mind or personality over living nature.'[12] Cope also believed that similar species develop in parallel. That they looked alike was only due to the fact that they followed similar evolutionary trends and environmental conditions, not that they had diverged from one another recently. They might share a common ancestor, but only in the distant past. His was not a branching tree of life but a series of parallel lines rarely touching one another.[13] Through a delicate mix of Bathmisms and Archaesthetisms variations were transferred to the embryonic form ready to be passed on to the next generation.[14] Therefore, both divine intervention and individual effort contributed to evolution. The internal mechanism was an evolutionary course correction from the Creator to get a species to go in the direction he wanted it to go. The other mechanical systems then functioned as a response to this cosmic nudge. Once nudged in a certain direction, the species stayed on that path, carried along by biological momentum. Later, Cope modified his theory somewhat. In his more flexible version, evolution was less a result of the touch of God and a cosmic mindset, and more the direct actions of the environment and utilitarian, Lamarckian-acquired characters.

This theory's structure may be a result of Cope's Quaker background. Quakerism believed in a transcendent God who enters man's consciousness and guides him along a predetermined path. Cope's metaphysics was surprisingly like that of several philosophers of the era, from Ralph Waldo Emerson to Madame Blavatsky. Cope argued, as they did, that there was a great universal mind of which all lesser minds are a part. Therefore the great mind of God can influence the course of all the others. Cope believed evolution was progressive: from simple to complex; lower to higher; imperfect to perfect. The end result of this was an enlightened spiritual consciousness epitomized by man. Peter Bowler argues that Cope did not want to be ousted from the scientific community and so played down the supernatural elements in his work for fear of being ridiculed.[15]

Osborn and Evolution

Cope saw Osborn as an ally and kindred spirit. By the 1880s or 1890s their relationship had grown beyond the student/mentor stage into a genuine friendship. They shared many ideas about evolution and its mystical elements, and were in agreement over the role of the divine. Although Cope's intellectual star was in decline, following the débâcles with the United States Geological Survey and his investment failures, Osborn's was in the ascendancy. Osborn found echoes of much of his evolutionary belief, particularly the mystical parts, in Cope's Bathmism or internal guiding principle, as well as Darwin's struggle elements, which he would later employ in his own work.

Osborn began writing evolutionary articles in the late 1880s. Following Cope's lead, he centered his studies on fossil teeth. The evolution of teeth, especially horse teeth, was important because in the changes in tooth structure Osborn saw what he believed was the slow, steady, progressive march of evolution. He believed that parts of an organism went from simple unspecialized forms to more specialized varieties as time passed in a series of slow incremental changes. These changes were brought about by the use and disuse of parts, occurred at the points of most wear, were due to the organism's environment and lifestyle, and were passed on to the next generation in a linear fashion. As Osborn put it, 'these new [tooth] cusps arise in perfectly definite lines of growth.'[16] He argued that every discovery in primitive mammal fossils 'tends to confirm the theory' that molars had gone through stages beginning with the simple single reptilian, or homodont cone, and proceeding to the more complex tritubercular (multi-cone) stage which appears in most lower Eocene mammals, and on which Cope did such pioneering work.[17]

He was not yet convinced, however, of just how organic change took place. The evolution of these teeth might be due to 'accidental variation' or use and disuse of parts, Osborn said; both theories had strong arguments in their favor. Though he had reservations all round he tended to shy away from the suggestion that anything happened accidentally. Studying variations in fossil teeth convinced him that at least some hereditary characters followed 'fundamental predispositions' in their development.[18] He saw the tritubercular molars as superior to the non-tritubercular forms (which he called 'degenerate and aberrant'), since they persisted into later ages.[19] The trituberculars did have their drawbacks. As their specialization and complexity increased, so did their chances for getting into trouble. Osborn believed that as an organism moved away from the general it was less able to adapt to environmental change. Therefore a less specialized, more general type would last longer and so was superior. Wanting to build up evidence to support this idea, he scoured fossil collections whenever and wherever he could find them.

In the summer of 1888 Osborn traveled to England to consult fossil collections there and give a talk at the annual meeting of the British Association for the Advancement of Science, at Bath. There he encountered Othniel Charles Marsh. He was annoyed with Marsh because of his work on fossil horse teeth. Marsh's collection was spectacular and growing even larger due to the efforts of collectors like John Bell Hatcher. This collection would allow Marsh, in theory at least, to make great strides in the study of evolutionary trends in horses, and thus mammals as a group. Osborn, however, considered Marsh's work to be a squandering of a precious resource since Marsh considered many of the teeth to be representative of different species. Osborn grimly told Wick Scott, 'every tooth is a new type of Genus. Oh it is rich!'[20] Osborn saw these slight variations as an indication of an organism's reaction to internal and external pressures. Different parts of the organism would react at different rates. These changes, he would later say, were not 'a sign-post on which to hang a new species [like Marsh thought] but as suggestive of the dawn of a new character.'[21]

Osborn and Scott's distaste for the man from New Haven was only getting stronger as the years went by. They would while away long lunches in Princeton dining rooms plotting, fantasizing and clinking their glasses over Marsh's downfall. Marsh in his pomposity exasperated Osborn further that summer in England. At the meeting in Bath, Osborn was scheduled to follow Marsh in the line-up to discuss his work on fossil horse teeth. Marsh went long over his scheduled time (forcing Osborn to cut short his lecture) and anticipated Osborn's talk.

While there are exciting group activities in paleontology, like field expeditions and such, much of the science was, and still is, done by sitting for long hours by oneself in dusty, dark storerooms of libraries and museums measuring and examining fossils. The slight by Marsh made Osborn angry and heightened his sense of loneliness for his wife and family. After the hustle and bustle of the meeting and the encounter with Marsh he told Loulu that 'I felt a thrill of longing to hold you tight in my bosom again and tell you of my love … and darling if you only knew what intense loneliness I suffer when all the excitement is over.'[22] Trying to ignore Marsh as much as possible, he completed his work on the fossil teeth, packed up his notes, and took the next boat home.

Henry Fairfield Osborn delighted in his family. His wife and children were the joys of his life and constantly in his thoughts, especially since his work took him away from them so often. Both La Casita in Princeton and the Garrisons estate in New York reverberated with their delights and upheavals (his mother Virginia helping out with the grandchildren), and he loved it all. While Loulu was at Castle Rock with the children he could concentrate on his work, but still found time to be part of their lives. Once, a child's prized shoe had gone missing and Osborn scoured the townhouse in search of it. He found it, like a piece of buried treasure, at the bottom of a toy box and shipped it upstate with a note proudly proclaiming he had found it.[23] Preoccupation with these small pleasures helped him keep his mind off weightier subjects. He was still troubled over the deaths of his siblings, particularly his sister Virginia, and in 1887 he wrote his mother to discuss it. She was moved that he remembered the anniversary of that tragic summer, since apparently no one else did. He would have more tragedy to fill his thoughts, since the deaths of close friends and relatives were never far from his door during the 1890s. He was also beginning to ponder the mechanisms of evolution more deeply, and to consider that they might have some bearing on more lofty questions. At the same time, however, he began to question the evolution theories of his colleagues, extending his doubts from the ideas of Darwin to those of Edward Drinker Cope.

One of the first major developements Osborn had to incorporate into his work was the germ-plasm theory of August Weismann. Osborn respected the work of Weismann, but he could not (as Cope and the other neo-Lamarckians could not) accept it as proof of natural selection. He distrusted random variation and natural selection because they suggested that life arose from chaos, and the absence of direction and a Creator. Osborn liked Weismann's theory in outline, but feared the neo-Darwinians were taking it in a 'radical direction he [Weismann] neither foresaw nor advocated.'[24] Osborn saw the germ-plasm theory as 'eminently conservative,' and saw no proof that the somatic cells were segregated from the germ plasm as Weismann said they were. If the cells were not separate, then changes acquired by the body could migrate to the germ plasm and be passed on to offspring. He saw 'internal reactions' at work, but not the type that Weismann envisioned. Osborn said that fossil evidence proved direction in evolution and that changes seen were 'reactions produced in the individual … [brought on by environmental conditions],' and that 'we infer these reactions are transmitted.'[25]

In 'The Paleontological Evidence for the Transmission of Acquired Characters' (1890), Osborn introduced many of the themes that would dominate his work for the rest of his life. Though steeped in the neo-Lamarckian tradition, he was not ready to accept the concept completely. Osborn began to argue a variation on the theme of both Lamarckian and Darwinian evolution in which incremental changes were transferred and added to permanently existing characters in the germ plasm. Osborn referred to his hybrid as 'Lamarckism *plus* – not *versus* Darwinism [his italics].'[26] An internal growth force in combination with changes in the environment and the organism's struggle to overcome those changes would trigger these characters into action and produce structural adaptation.

This passing on of tiny variations was clearly illustrated, he said, by the fossil record. Osborn was, however, growing increasingly dissatisfied with current evolutionary thinking. He was not sure either school really explained organic change. He acknowledged the feud between the two sides, but saw it as counterproductive since it made the scientific community seem fragmented. He was torn between neo-Lamarckism and neo-Darwinism and wrestled with the evidence in his papers and correspondence. In his writings we can see him struggling to figure out which side to choose. He told a friend that 'for myself in a vast number of characters I can see no cause except selection – while in another class I can see no selection at all.'[27]

The friend to whom Osborn made the natural selection comment was British biologist Edward Poulton. They met studying embryology in London and became friends. Poulton was more of a Darwinian, but their mutual respect led to a lifelong relationship where they continually argued the merits of various aspects of evolution. Shortly after August Weismann's germ-plasm theory went public the two friends talked about it. Osborn was at first skeptical (he had not yet read the German's work entirely) and since he believed in the direct action of the environment, looked forward to the germ theory's demise. He told Poulton that 'Cope rubs his hands in glee, at the prospective feast upon the Weisminians.'[28] Osborn's elation that the neo-Lamarkians would trounce the neo-Darwinians over Weismann was short-lived. Having examined the material more closely, he hesitated. A few months later he told Poulton rather sheepishly that 'I am postponing my discussion of the acquired characters question until I see Weismann's theory as a whole.'[29] Though a bit chastened, he still enjoyed his intellectual sparring with Poulton, feeling it kept him thinking and alert. Their conversations also showed Osborn the 'weak spots' in Cope's work. He admitted to Poulton that 'I am not wedded to my theory [neo-Lamarckism] and will not be heartbroken if it does not survive two more years,' and added: 'Weismann makes the transmission of acquired characters almost an impossibility.'[30] Equally dissatisfied with Darwin, he said, 'the more I think of it, the more inadequate natural selection theory seems to me to explain evolution.'[31] He challenged Poulton to look at the fossil horse series and tell him if he could account for it by natural selection. If he could, Osborn told him, he would accept it.[32] Poulton did not have to work that hard. Osborn was already slipping away from Lamarck and more towards natural selection all by himself. The appearance of Weismann's theory and its seeming validity only made Osborn doubt Cope's work more.

Having now become intrigued by the germ-plasm theory, Osborn modified it to make it more satisfying to his tastes. He said the hereditary basis of any species, or 'race,' was a core of foundational characters which accumulated very slowly and steadily from the deep past. While some characters were acquired later, maybe even by chance, they had little overall effect. The important elements were those in what he called the 'race plasm.' The race plasm formed a continuous chain to the past which was 'in a planetary sense, "immortal."'[33] An individual organism derived its characteristics from this permanent ancestral pool. He believed that ontogenetic variations (those occurring in the individual) were at best temporary: they came and went. Phylogenetic variations (those affecting the entire group or phyla) were what lasted. Therefore, the evolution of early forms dictated the evolution of newer ones because of this built-in racial hereditary foundation, the result of which would be 'an appearance but not the real transmission of aquired characters.'[34]

Osborn's theory of the relationship between ontogenetic and phylogenetic change that did not rely on Lamarckian mechanics became the core of his thinking. He claimed that 'centers' of variation, such as the teeth and jaws, 'seem to have predetermined at what points certain new structures would appear.'[35] Whereas Darwin saw evolution as mechanical (two independent processes – variation and environmental factors – producing random adaptation), and not the result of some intrinsic life force, Osborn saw it as a materialistic process which was the result of an internal force that allowed, or even prompted, the organism to evolve. Environment played its role at this crucial point. Changes in the environment would trigger built-in characters to action. Once the germ plasm reacted to the environment, and started down its predetermined path, however, it did so regardless of the vagaries of the environment.

In addition to the work of Weismann, Osborn used that of Darwin's cousin, Francis Galton, to support his contention that there was a fixed base hereditary pool of ancestral, racial characters. Galton showed, according to Osborn, that 'individual variations are conspicuous but in the main temporary, while there is a strong undercurrent of phylogenetic variations relatively inconspicuous and permanent.'[36] This condition only prevailed, however, within the 'natural inter-breeding' of species. Osborn would argue that unnatural miscegenation was deleterious because it would upset the permanent characters and throw evolution into turmoil; hence ancient species never interbred, nor should modern ones. The distinction between natural and unnatural breeding was an idea that would play a crucial role in Osborn's later thinking about human evolution, science and society.

Osborn claimed fossil molars showed that the teeth changed to meet the needs of a changing environment, and that the modifications were passed on to the next generation in a linear fashion. His favorite example, which became his paradigm of mammalian evolution, was the horse. (Hundreds of often-complete horse fossil skeletons had been found over the years, representing every variation from the earliest small four-toed *Eohippus* to the larger three-toed *Merychippus* and *Neohipparion*, to the modern one-toed *Equus*). Different species of horse, for example, evolved similar tooth structures because they inhabited the same type of environment which was exerting similar pressures, not because one group gave rise to the other. Osborn argued that since horses 'of the Rocky Mountains, the Thames Valley, and Switzerland!' all show the same adaptations, random variation and natural selection could not possibly be the driving force of evolution, and that species evolved in a parallel fashion, not a branching tree. 'Evolution is one of uniform and uninterrupted progression' which had purposeful direction, and thus was definitely not random. Mammal groups that seemed to have risen from one another had diverged from a common ancestor not recently, but in the distant past. This divergence occurred, he argued, so far in the past that by the time the groups spread out, they were no longer directly related.[37]

Osborn's version of evolution was a complex interaction of external and internal forces working along a path in a process he called 'organic selection.' However, when he tried to publish the idea as a new breakthrough in evolutionary thinking he was scooped by James Baldwin and Conwy Lloyd Morgan and their theory, also called 'organic selection.'[38] Their work, which attempted to do something similar to his, supported the neo-Darwinians, and since they were able to publish almost a year before him, theirs eclipsed his and organic selection came to be known as the Baldwin Effect.[39]

In the last decade of the century, Osborn was still torn between Lamarck and Darwin, and continued to list the advantages and disadvantages of both theories, comparing them and modifying them to suit his needs as he went. He continued to see elements in both which were useful to him, and this conflict never allowed him clearly to choose one theory over the other. Though he modified Weismann and Darwin to suit his aims, and finally acknowledged that Lamarckian thinking had been proved false by Weismann, Osborn's work retained a Lamarckian flavor that he could never shake off. Osborn's intellectual struggle was a human drama played out in his articles and correspondence as he tried to reconcile his personal beliefs with the facts and theories around him. He bounced ideas off of Poulton through personal correspondence, and with the scientific community through his articles. Simple empiricism was not all that was guiding him. Seeing evolution working as a result of the same type of tensions and turmoils that affected not only existence in general but his own in particular, he said, 'life is a continuous adjustment between internal and external relations.'[40] He would need all his ability to juggle those internal and external relations if his theory were to survive.

The New Biology

Not only were changes occurring in Osborn's thinking on evolution, but in his personal life as well. By 1890 his father William Henry Osborn had entered the final phase of his life. He was almost always bed-ridden after having suffered from a brain seizure and was fading quickly. In 1891 Henry Fairfield Osborn was offered jobs at both Columbia College (to set up the new biology department) and the American Museum of Natural History in New York (as head of the mammalian paleontology department). Though he loved Princeton dearly, he never thought he had been treated quite the way he should have been by the administration. With Scylla now an ally, and Charybdis dwindling at Yale, he thought it safe to take the job at the American Museum he once turned down. It was also not such a bad idea to leave the relative obscurity of Princeton to begin carving out an empire in Manhattan, and so he took the family north and bought a townhouse at 850 Madison Avenue. Both institutions wanted Osborn for his political connections as much as for his scientific acumen. The tribal elders of New York society, who were the prime benefactors of museum and cultural life there, felt comfortable with such a man as Osborn since he would understand their desires for social order and stability and be willing to fulfill and promote them. They wanted to instill in the populace that science and technology were important. They also wanted museums to teach moral as well as technical lessons. They wanted museums to teach people 'the higher ideals which contemplation of nature was thought to encourage.'[41]

The double duty put extra stress on Osborn and made his life more complex than it had ever been before. As a way to relax, he took up bicycle riding. He especially enjoyed it because his son Perry would often accompany him when he rode to work at the museum or the school. Once at the museum little Perry would play in the halls while his father worked. He also took the children on field trips as well as on vacation. During a trip to the Woods Hole Oceanographic Institute in Massachusetts to give an address, the children learned to sail and swim. Henry Fairfield told Loulu that involving them was 'really great fun!'[42] He would need all the fun he could get. While his professional life was looking brighter and brighter, his spiritual life was about to enter a particularly dark phase.

Osborn began creating the biology department at Columbia at a time of great change for the discipline. It was the advent of what was called the 'new biology,' a period in which nineteenth-century morphological techniques and thinking were giving way to the laboratory experimentalism of the twentieth century. An important aspect of what characterized the new biology was a philosophical attitude about where and how science should be studied. Historian Garland Allen argues that the younger men had two main complaints about the older morphologists. First, the new men saw the old ones as too rigid in their thinking, seeing all questions of morphology and form as related to evolution only, and neglecting important questions of heredity, embryology and function. 'It was felt that the study of form had totally eclipsed the study of function.'[43] The other was that the older morphologists were too speculative in their broad generalizations and were creating theories that could not be tested or proven one way or the other. They argued that mounds of theories, any one of which might be valid, brought you no closer to a firm understanding of how the universe worked. By the 1890s younger biologists 'had become exasperated with the aims and practices of morphology.'[44] They wanted to open up biology and study other than directly related evolutionary questions for their own sake: questions of heredity and cell mechanics, for example. In their laboratories the younger men could manipulate cells and embryos with an increasingly sophisticated battery of techniques, ask specific questions and look for specific answers, using more deductive approaches. The older morphologists were often inductivists who built up observational evidence, allowing nature to lead them wherever it wanted to go. On the other hand the older men accused the younger men of setting up artificial situations in their sterile laboratories that did not exist in nature. Therefore, any results gathered through this artificial means were faulty.

Ronald Rainger, Keith Benson and Jane Maienschein, in *The American Development of Biology* (1988), argue that Allen is wrong in his assessment of the old versus new biology debate. Though they do acknowledge momentous changes going on in the biological profession, they see no clear break between opposing forces. The biological sciences, like American society in general, was going through a period of self-definition. Biology in the United States had begun to 'develop a distinctive American character.'[45] Osborn clearly saw a distinction between the older naturalists and the younger experimentalists and called it a 'fratricidal conflict.'[46] This division added to his inner turmoil. Though raised a morphologist of the old school, he was astute enough to recognize the importance and usefulness of the new approaches. He wanted to go forward, but was held back by his own agenda. He tended to oppose the experimentalists because some of their work went against his Victorian naturalist's beliefs about how science should be done. The old naturalists spent their time in the great outdoors studying the book of nature directly. The new experimentalists did their studying in a laboratory, reading the books of men, not following Mother Nature's rules, but making them up for themselves. Osborn recoiled from this seeming abandonment of nature to the sterile walls of academia and revolt from established patterns. While he had no objection to experimental lab work in itself, he felt the way to study the natural world was by observing it first hand. The increasing intellectual specialization of the era was a dangerous trend to him because being locked up in a laboratory kept the student from gaining the complete picture. No accurate analysis could be made if the student did not get his feet wet in the field. Just as overspecialization could lead to the extinction of living forms because they could not adapt creatively, this too would be the fate of science if it became too rigid and focused. Total reliance on experimentalism and hyper specialization seemed to him unnatural. It broke the natural order of things, divorced man from nature, and led to a questioning of natural authority – none of which was good for science or society. He could not, however, deny that the new men were on to something.

Osborn's indecisive position can also be seen in his relationship with Thomas Hunt Morgan, the man who came to epitomize the new experimental approach. Morgan had pioneered the growing discipline of genetics with his celebrated experiments on fruit flies done at Columbia University in the early part of the century, making crucial discoveries in mutation and heredity. Although at first skeptical of Darwinism, Morgan's experiments eventually converted him. He 'sought to make genetics the foundation for exploring development and evolution.'[47] Osborn, as head of Columbia's biology department, insisted on bringing Morgan in as a way to turn the fledgling office into a world-class institution. In 1901 Osborn had helped correct proofs of Morgan's *Experimental Evolution*, and in 1903 he told the board of the college he would give up part of his own salary to insure Morgan's hiring.[48] Despite this support, once Morgan began his experimental program Osborn became uncomfortable with it. He thought the fruit-fly experiments were done 'under unnatural conditions.'[49]

It would be easy to say Osborn regretted hiring Morgan, that he was disappointed that Morgan did not help him discredit strict neo-Darwinism, and that Morgan turned on him. Morgan ridiculed Osborn's idiosyncratic use of language, which he saw at best as misrepresentation and at worst as outright obfuscation.[50] He saw Osborn's books as examples of the 'worst features of speculative theory,' and knocked paleontologists because their minds had a tendency 'to wander.'[51] However, Osborn was always complimentary to Morgan, and there is no evidence that he tried to get rid of him. Indeed, Osborn called Morgan 'one of the leading experimental zoologists … the author of many important works.'[52]

There was more than just this intellectual tug of war to occupy Osborn's mind. The year 1894 saw two momentous events in his life. He had invited Wick Scott and Edward Drinker Cope to New York to have dinner and discuss evolution on the first of March that year. He had finished the manuscript for his first book and was eagerly awaiting its publication in August. *From the Greeks to Darwin* was a history of evolutionary thought from the classical age to the mid-nineteenth century (though artfully avoiding the controversy over natural selection). The dinner never took place because his father took a turn for the worse. The old man was finding it difficult to recognize friends and family members. When he did finally recognize an old acquaintance, William Henry Osborn burst into tears at the realization of his own condition. The sadness of it all made Henry Fairfield think of his own family. While relating the scene to Loulu he stopped in mid-thought and told her, 'Do be lovely, but not indulgent with the children.'[53] With his wife Virginia and most of the family around him, as he lay in bed at 32 Madison Avenue, William Henry Osborn, the rock which built the Illinois Central Railroad, taught his children the importance of striving to achieve, and who kept Princeton University from going under, died quietly on March 2, 1894 of the illness which had dogged him most of his life.[54] Condolences came in from sources as diverse as the eccentric politician Theodore Roosevelt, and the eccentric scientist Nichola Tesla. In an additional moment of poignancy William Henry's old Princeton ally Reverend James McCosh wrote, in a shaky hand which heralded his own poor condition, that he would soon be joining his old friend in the great beyond – which he did the following November.[55]

His father's death drew Osborn to thoughts of mortality. Quoting from a novel he read during his father's last days, he talked about 'the surrendered will' (his underline). It was something reached through work, duty and pain. This will is 'the root, the source,' which 'leads us in life and carries us in death.'[56] The following August grandmother Sturgis died. Virginia Sturgis Osborn had prayed to God to let her mother pass quietly, and according to her He granted the wish.[57]

The pain of his father's death was offset by the birth of his fifth child, Gurdon Saltonstall Osborn, on June 14, 1895. The elation over this newest addition to the Osborn clan did not last long. It soon became apparent that little Gurdon was not a healthy baby. He had no appetite and, like William Henry Osborn, slipped in and out of good health. After a year of emotional ups and downs over the baby's condition (and shortly after the anniversary of William Henry's passing) the baby died on March 13, 1896. (The death of the child may have slowed Osborn's work enough to postpone his publication of the organic selection idea, allowing J.M. Baldwin to appear with it first.)

Henry Fairfield's brother Willie had been on an extended tour of Europe when the child died. He had been musing over the death of their father at the graves of Shelley and Keats when it happened, and taking in a dark and somber Catholic Mass.[58] He shared his brother's grief and told him his thoughts were with him. Henry Fairfield took to his bicycle to pedal away his feelings. His mother Virginia had also been out of the country, visiting the Holy Land as a way of dealing with the loss of her husband and her mother. Upon her return she went to spend time with the Vanderbilts at their palatial estate, Biltmore. Between 1894 and 1897 Henry Fairfield Osborn experienced the deaths of his father and his own child, as well as those of several other people who were close to him. Hard work helped him cope with it, and during this period he was closing in on ideas which would set momentous wheels in motion and set his intellectual life path.

The Death of Cope

Henry Fairfield Osborn was a prolific writer who believed that scientific knowledge should be accessible to the lay population and argued that a well-educated public (well educated in science particularly) was good for the country. Appreciation of the laws of nature fostered appreciation of the laws of man and the importance of obeying those laws. Nature study was a socially redeeming practice that reinforced social order.[59] To that end, in 1891, he published an article on 'The Present Problem of Heredity' in the magazine *Atlantic Monthly*.[60] It would be the first of a long line of articles and books meant specifically for popular audiences. He saw educating the public as a solemn duty – which also benefited himself by bringing the public around to his way of thinking – and he utilized any magazine which gave him space. One of his favorite vehicles for teaching the public how it should think was *Century Magazine*.

A significant aid in the effort to spread the gospel of Osborn came along in 1896 in the form of the young New York illustrator, Charles R. Knight. Knight had already done work for *Century* and other magazines, and was ready to add more clients to his list. He had also become known around the American Museum because of his extensive use of the museum's exhibits and holdings as subjects for drawing and learning about the natural world. After making the acquaintance of museum paleontologist Jacob Wortman, Knight was introduced to Osborn. They shared views on art, politics and the state of American society and so quickly became friends. It was the beginning of a relationship that would mold and dominate the public's perception of what dinosaurs and cavemen looked like, one which lasted throughout most of the twentieth century. As a man who saw public education as a way to promote several of his agendas, Osborn knew that a well-crafted picture could be worth thousands of words.[61] Both Cope and Osborn were distressed over the sorry state of representing prehistoric animals that had not progressed much beyond the pioneering, but comical work of British artist Waterhouse Hawkins and his Victorian monsters.[62] Osborn introduced Knight to Edward Drinker Cope, knowing that the mad genius of Philadelphia had just overseen an article for *Century* called 'Strange Creatures of the Past' written by William H. Ballou. Since his childhood days studying ichthyosaurs Cope had been making carefully detailed drawings. He supplied Knight with several rough sketches and plenty of technical information on how to draw dinosaurs properly. Knight spent time with Cope at his Pine Street residence in Philadelphia in an intense period of study. Unfortunately for Knight this relationship ended just as it was beginning.

Edward Drinker Cope and Henry Fairfield Osborn had grown very close over the years. Osborn had come to know the stormy Quaker as well as anyone could. Like characters in a Greek drama, Cope was the wise old sage who pointed the young hero in the right direction, imparted secret knowledge to him, and gave him the weapons needed to defeat his foes. Once a master himself, Osborn returned to save the dying king. Osborn had great respect for Cope and, while sometimes disagreeing with him, or questioning his behavior, never for a moment gave up his loyalty and respect even when the scientific world passed Cope by and pushed him offstage, to be forgotten. Osborn's biography, *Cope: Master Naturalist* (1931), is in large part an attempt to resurrect Cope's memory, paint him as the victim in the bone wars with O.C. Marsh, and generally show Cope to have been a man of good humor and affection, not the bitter, angry eccentric most thought him.[63]

As early as 1886 Osborn was coming to Cope's aid. He had sided with Cope in the fracas with Marsh and the United States Geological Survey, and had tried unsuccessfully to get legislators in Washington to give Cope back his federal funding. Cope tried to get Osborn to buy his journal, *The American Naturalist.* With an edge of desperation he told him, 'You can get it easily!'[64] The journal was in deep financial trouble, as was Cope himself, and he hoped the younger man would bail him out. Osborn ended up giving Cope $2,500.00 to cover that year's deficit and keep both man and journal afloat. Osborn had several motives. First, he genuinely wanted to help Cope, and second, the journal provided a ready source of publication for Osborn's more scholarly articles and ideas. Cope felt lonely and abandoned by his profession, and looked to Osborn for professional as well as personal support. In order to help out his friend, and secure a magnificent collection for himself, Osborn had the American Museum buy Cope's North American mammal fossils for $32,149.89 in 1895.[65] While he was sorry to lose such a valuable collection, Cope needed the money to support his family. He had been reduced from leadership in the American scientific community to begging for scraps from his former student.

By March of 1897, however, Cope was gravely ill, 'perhaps fatally so' with 'cystitus' (*sic*).[66] The years of manic work in the wastes of the Western territories, squabbles with colleagues and poor health habits were coming to a head. Cope tried drastic measures to cure himself, including taking the preservative formalin, something which alarmed Osborn. He tried to get him to see a doctor, or at least take better care of himself, but to no avail. He even set up an appointment with a New York hospital for tests and surgery, but Cope did not go, opting instead to take 'an enormous quantity of morphia and finally resorted to belladonna.'[67] With great trepidation Osborn went to see the dying king one last time on April 11. The visit was not encouraging. Osborn had just buried his father and realized that he would soon have another funeral to attend. Back at the museum he wrote Wick Scott to say that 'I left the death bed of poor Cope this afternoon.'[68] Struggling to the end (he had insisted on teaching his classes that week even while he teetered on the edge of delirium), Cope died the next morning. As Cope's executor,

Osborn presided over the simple, quiet Quaker funeral held at the Pine Street house. The ceremony (there were actually two, one for colleagues, one for family) consisted of a few close friends, including Wick Scott, sitting in Cope's beloved library with his casket propped up on chairs. Lining the walls and covering the floors of the room were piles of fossils Cope would never get to examine, and books he would never read. When Osborn finally stood to speak, he quoted Job. 'Where wast thou when I laid the foundations of the Earth … and all the sons of God shouted for joy?'[69] The only reply was the ticking of the clock and the forlorn scratching of Cope's pet turtle and Gila Monster as they searched in vain for their master. With that, the last great American naturalist was gone.

Further Complications

By 1887, as Ronald Rainger points out, the basic framework of Osborn's theory of evolution was complete. He now had to defend it from the onslaught of new theories that threw his work into question, a position that would not be easy to hold. The work of August Weismann was not the only hurdle he needed to overcome. By 1900 the Dutch botanist Hugo de Vries had resurrected the forgotten work of Gregor Mendel, and helped show mutation to be at the heart of variation. Mendel, a Moravian monk, had done a series of experiments with pea plants in the 1850s and 1860s. His results led him to theorize that the male and female each contributed hereditary material which combined in a certain ratio of dominant and recessive traits, creating the physical characteristics of their offspring. His work was published in an obscure journal and not 'discovered' by the scientific community until 1900. Mutations in the germ plasm as a basic idea was acceptable to Osborn, but De Vries called for large-scale random mutations that could produce a new species in a single generation. Osborn naturally bristled at the notion of a random process and saw no evidence in the fossil record to support species appearing that quickly.[70] In addition, De Vries's methodology was squarely in the experimentalist camp. To avoid this dilemma, Osborn attempted to place himself in a position where he would not have to accept or repudiate De Vries or the growing trend of new biology. He was already beginning to feel left behind. He told Wick Scott in 1903: 'I think it is necessary to restate the arguments for the Tritubercular theory because it is otherwise in danger of being thrown over altogether.'[71]

Before the appearance of the work of De Vries and Mendel, Osborn had become interested in the mutation theory of German invertebrate paleontologist Wilhelm Waagan (1841–1900). William Berryman Scott had begun promoting Waagan's work in the early 1890s and Osborn picked up on it. Waagan, according to Scott, saw external forces working on an internal guiding force.

The resulting 'mutations' built up slowly until they produced organic change.[72] Waagan formulated his views based on the fossil record of ammonites (a marine group whose closely successive fossil series undoubtedly reminded Osborn of the horse series). These observations led Waagen to believe that a process (later called *Mutationsrichtung* by another German, Melchior Neumayr) caused forms to evolve in a definite direction. Waagan's mutations were slow and continuous, the way Osborn preferred them, while De Vries's were rapid and discontinuous. Osborn claimed Waagan's theory showed for the first time how a new species was seen to rise in nature.[73]

As was often the case, Osborn was not totally opposed to new theories and as such tried to finesse the mutation question. He referred to the rediscovery of Mendel's work as 'the greatest biological discovery of the year and perhaps of the decade.'[74] He tried to get around De Vries by saying that a paleontologist could only make decisions based upon the fossil record. Thinking himself safely above the fray, he continued that he was 'neither opposing nor advocating the so-called "mutation theory"' of De Vries. He insisted: 'I simply assert that paleontology is not a branch of biology in which this theory can be either proved or disproved.'[75] Since everyone else seemed to have 'eulogized De Vries to the skies', he felt he should play devil's advocate. He told Loulu that his own theory 'represented a strong argument against the wide scope which the learned Dutch botanist had given his theory.'[76] Feeling the tug in both directions, Osborn later gave in and admitted that De Vries's mutations were acceptable as long as he could argue they supported predisposed hereditary trends.[77]

As scientific advances progressed he kept dancing around them and finessing them as need be. He gave up ground as long as he could keep the core of his belief – even reaching a point where he conceded that other organisms might evolve according to neo-Darwinian principles but not mammals, as they were special. He had to hold on to the mammals or he could not deal with human evolution the way he wanted. He said that 'mutation plays a relatively small part in the evolution of mammals.'[78] Osborn's selectivity regarding which groups evolved in which way prompted T.H. Morgan to chide him, asking if he could have the mammals back into the family of living things when Osborn was done with them. In response, Osborn posed the same challenge to Morgan that he had to Edward Poulton. 'If you will,' he quipped to Morgan, 'be good enough to point out any single important structure … which has arisen through mutation, I will be glad to let you have the whole class [back] within the fold of the faithful.'[79]

Here we see Osborn's conflicted side coming through again, trying to hold on to a past he knew to be tenuous, while tentatively reaching out for a future he was afraid of. By combining theories or co-opting them, Henry Fairfield Osborn was able to formulate a bulwark against Darwinian variation and the theories of the experimentalists as well as to provide a foundation for his theory of human evolution and his views on modern society. Yet he was continuously giving ground to those very theories he said he opposed. He had abandoned neo-Lamarckism while still keeping the parts he could use. He had accepted the work of Weismann by claiming it really showed an internal perfecting and direction-producing process. Having adopted Waagan's mutation theory, he had a tool to combat De Vries's theory before it appeared.

He was eventually able to accept natural selection because he had convinced himself that a non-random process was at work in the germ plasm. Allowing in more and more Darwinian thinking (at least his version of it), he hesitated to give in completely. He stated flatly: 'I am not a Darwinian … I do not think we have solved the evolution problem. Perhaps we never shall, but the survival of the fittest (his underline) is a law as well demonstrated as the laws of gravity and this is the essence of Darwinian teaching.'[80] Natural selection could have its place since directed germ-plasm developments were naturally selected to travel in directed paths to predetermined outcomes. And just in case anyone did not understand that the ultimate workings of the germ were unknowable, or think experimentation might be able to explain it, Osborn said that 'Germ [plasm] evolution is the most incomprehensible phenomena which has yet been discovered in the universe, for the greater part of what we see in animal and plant form is only the visible expression of the invisible evolution of the hereditary germ.'[81]

He had others searching for invisible elements as well. While working with explorer William Beebe – who was then in British Guyana setting up the American Museum's Tropical Research Station – Osborn told him to do extensive experimentation on the life forms there in order to advance his theory of 'unknown' evolutionary agencies. He told Beebe not to forget that the most important aspect of science was the accumulation of 'facts – facts – facts [which will] crystallize into generalizations.'[82] Trying to combine Victorian induction and modern experimental deduction, Osborn wanted to show that chemical and mechanical adaptation were separate. He wanted Beebe to look, not just at the usual known agencies 'such as light, heat, gravitation' and their role in development, but also at 'unknown agencies such as vibration.' He was convinced that 'there are sense organs of which we have not thus far dreamt.'[83] Just what sort of vibrations or sense organs he meant, he was a little vague about. Osborn was fascinated by the idea of these nebulous processes at work in the universe which he felt were an indication of the hand of God.

Though his latest child – Gurdon Saltonstall – was dead, the other Osborn children were growing up quickly. Henry Fairfield Jr – known as Fairboy – went to the exclusive Groton preparatory school in Connecticut along with Theodore Roosevelt Jr, but was having a rough time, rating only a B– average. By 1904, however, he had whipped himself into shape and passed his exams.[84] These setbacks could not dampen his father's enthusiasm, though.

In 1905, on the boy's eighteenth birthday, he beamed to Loulu, 'what a lovely present you made me eighteen years ago this evening.'[85] He still had an eye for her as well, marveling at her 'buxom appearance,' and was glad she had 'no skeletons in the closet.'[86] While he loved his wife and all his children dearly, Henry Fairfield Osborn held a special place in his heart for his youngest daughter, Josephine. She had been raised a little princess, but she was also expected to use her mind, her talents, and natural gifts and advantages to advance herself in life. She received music lessons, formal schooling and many trips abroad, where she became the toast of England and the continent as a socialite. She and her father had a close relationship and often confided in one another about how they were feeling. When Josephine worried whether or not she would become conceited and develop a superior attitude because of her background and abilities, he told her to hold herself high and set her standards even higher, 'so high … that you never quite reach it.'[87] This way she would remain humble and know that constant effort and struggle were the only way to succeed.

Her worries mimicked his own. He was afraid of becoming arrogant and developing a superior attitude. His worry over becoming big-headed seems genuine, though his self-deprecating manner at times seems insincere, as when he remarked: 'I am by nature so modest – one of my greatest charms by the way.'[88] Concerned over a portrait being painted of him, he told Loulu, 'it makes me as handsome as you think I am – and scholarly looking as I ought to be.'[89] He was blissfully unaware of just how conceited and arrogant he was becoming, and how his reputation as a lofty self-important patrician was growing apace. He could be abrupt and dismissive to underlings and menial job holders. George Gaylord Simpson (1902–84), who worked for Osborn early in his career and who would later become a major contributor to evolutionary theory in his own right, remembered one incident shortly after the publication of one of Osborn's books in 1929, which became legendary. Osborn presented Simpson a copy just off the presses. When Simpson asked if he would be kind enough to autograph the copy, 'The Professor' did so with a dip pen and great flourish. As Simpson was about to blot the heavy wet ink Osborn stopped him, and with dead earnestness said, 'Never blot the signature of a great man.' Simpson liked Osborn and found his pompous behavior not annoying, as most did, but 'amusing and indeed rather endearing.'[90] Osborn always treated Simpson with respect and cordiality even when the younger man heatedly disagreed with his ideas. He knew that while Osborn could be imperious and vain, he rarely treated anyone poorly on purpose or out of intended meanness. In later years, when many in the scientific world, including many at the American Museum, publicly disagreed with Osborn's theories and even ridiculed him, he made few personal attacks, and never made his subordinates afraid to disagree with him for fear of being fired. Indeed, he encouraged them to disagree as a way of working out theoretical questions. Despite this, Osborn does seem to have been roundly disliked by many on the staff at the museum because of his attitude.

Aside from worrying about his attitude, Osborn was still busy as ever, working and traveling. He spent a good deal of 1905 and 1906 in the Badlands of Nebraska looking for fossil horses with paleontologist Walter Granger. Having just turned forty-nine years old he reflected on his life and saw that he did have a list of accomplishments to be proud of. From the West he wrote Loulu that she would be 'happy and proud of your man of science – who has made some sacrifices to accomplish great results and hold his head up among the other geologists and explorers.'[91] He then went to the Fayum Desert of Egypt as part of an expedition searching for the origins of the elephants. There he and Josephine visited the pyramids and got caught up in an ocean of Muslim pilgrims on their way to celebrate Ramadan. But he made sure to be home for his own favorite feast day of Christmas. He loved taking the family to church and filling several pews, especially during the holiday season, and had 'dear old 850 Madison Avenue' decked out in all its Victorian Christmas splendor. A few weeks later he and Loulu visited his old friend Theodore Roosevelt at his new home in the White House during the annual public opening of the Capitol where the president could greet citizens personally. Roosevelt listened in rapt attention as Osborn told him of the Fayum expedition. One can imagine Roosevelt grinning broadly and saying 'Bully, Osborn, bully!' A month after that, the board of trustees at the American Museum of Natural History gave him a belated Christmas gift by making Osborn himself president in a ceremony of religious solemnity that took place in the imposing personal library of his uncle J.P. Morgan.[92]

Predestination

One of the themes that linked Henry Fairfield Osborn's views on evolution and religion was predestination. Evolution, he said, went according to ways predetermined by the structure and heredity of an organism, which found its ultimate cause in the workings of the Creator. Protestantism in general had been moving away from strict Calvinist predestination. Osborn's mother, as well as Princeton's James McCosh, had reinforced in him the notion that predestination was not the strait-jacketing concept it had been in the past. This type of soft view of predestination had also worked its way into American mythology. That people could achieve more by struggling beyond their station in life was a powerful motivator at all levels of American society. Americans reveled in the belief that advance in any field was possible through hard work.[93] This more flexible religious view found its way into Osborn's science as well. As a result Osborn described hereditary characteristics in terms from religion and culture, saying that they were 'constitutionally predetermined.'[94] He believed that through struggle it was possible to break the bonds of built-in hereditary characteristics. He tried to show that an organism could exert some control over its life by personal will-power and determination. He was perilously close to being caught in the classic paradox of predestination versus free will. In biological terms this translated into the question of if an organism was moving along a predetermined evolutionary path (as Osborn said they were), how could its personal behavior mean anything? If there was one, how could there be the other?

This was one of the sticking points between Osborn and the new experimentalists. Many old-school morphologists saw the development of the embryo as something determined by previous generations. It formed in a fashion fixed by the past, unable to go beyond its pre-set path. The proponents of the new biology argued that the embryo was not a 'slave to its own prehistory,' it was not fixed by some ancestral predestination: it could undergo change.[95] Neither concept was completely acceptable to Osborn. He rejected total predestination, which would not allow for advancement, but neither did he embrace anarchic free will, which would throw the universe into chaos. He desired an artful combination of both.

The problem of predestination and free will was nothing new to American science. Botanist Asa Gray wrestled with it in the middle of the nineteenth century. Gray had corresponded with Darwin before 1859 and was sent a pre-production copy of the *Origin of Species* for comment. A New York Presbyterian like Osborn, Gray tried to find a teleological way to explain natural selection. For Gray and Osborn, as well as Cope and a host of other Victorian naturalists, God was the first indisputable cause of evolution, with non-random chance left out of the equation.[96] Not unlike William James later in the century, Gray had seen that the central issue of Darwinian evolution was the conflict between chance and design. Like Osborn, Gray was uncomfortable with the idea of a universe that had not been designed, or at least one which did not conform to divine law.[97]

By the turn of the century Osborn had rendered theological ideas into a scientific idea. He unconsciously equated the spiritual struggle for salvation with the physical struggle for evolutionary advancement. This secularization, blending science and religion, paralleled the way he hoped to blend Lamarckism and Darwinism. For Osborn evolutionarily successful species – those able to adapt to their changing environment and continue to breed – were superior. They had used their inherent characteristics yet struggled beyond them. Osborn had found a way to approach the problem of predestination and free will by using both to his advantage: not Lamarckism instead of Darwinism, but Lamarckism *plus* Darwinism.

In a tradition that ran from Augustine to Calvin to the Scottish Enlightenment to the Puritans, Presbyterians, and Congregationalists of New England, thinkers struggled to hold on to both free will and predestination, to reconcile providence and human behavior.[98] Students of evolution were the inheritors of this problem as expressed in the scientific questions surrounding whether or not a species could affect its evolution by its behavior. Though Osborn struggled mightily to do so, whether he could was still in doubt. He came closest to expressing it in 1908 when he introduced 'The Four Inseparable Factors of Evolution.' To Osborn, heredity, ontogeny, environment and selection were the building blocks of evolution. They worked together in a complex and shifting manner to produce evolutionary change. The key to this idea was that no one of the four was truly dominant. Different characters in an organism evolved when one of the four led the way. Sometimes heredity, sometimes ontogeny, might be responsible. When one took over the others followed. They were 'inseparable and interacting.' They all contribute but not always in the same order.[99] In this way he argued that some characters were predetermined and under hereditary control, while others were due to the organism's struggle with the environment.

As a naturalist of the Victorian persuasion, Osborn was enamored of the idea of discovering wide-ranging, generalized 'laws' to account for how nature worked. His four elements work together to produce change in a specific direction as determined by an interaction of these forces. He also considered this law the 'only true conception of the evolution process.'[100] Osborn did not include variation in his equation 'because it is a secondary process' and therefore not as critical to change. Bowler argues that Osborn included ontogeny in his four laws as a way of making 'a link between … earlier traditions … while at the same time he was clearly trying to make his position more acceptable in the light of modern knowledge.'[101] The whole of the organism was a combination of characters which were the result of predetermined evolutionary paths and personal struggle against the environment.[102] In the end, Osborn still felt that the predetermined hereditary element accounted for most of the evolutionary process, but he held out that a degree of change was the result of personal effort.

Conclusion

In the second decade of the century Henry Fairfield Osborn's chance to find the physical proof upon which to hang his entire idea began to seem within his grasp. Building upon his experiences with Guyot, McCosh and Huxley, with the influence of Cope, his work had matured and a basic framework of evolution studies had been set up. He had begun a lifelong wrestling match with the question of how humanity had evolved and, working backward, how and where it had originated. He would soon begin to apply the lessons he learned in nature and the fossil record about man's place in the universe to the practical questions of how civilization was and should be ordered. Though he followed a different path to get there, he wound up at the same place as Haeckel, Blavatsky and the other mystics and philosophers: Central Asia. Like the others, he had begun to consider questions of human racial qualities and their place in evolution. The Central Asian origins of man could, for him as for them, prove the specialness and exalted status of the white race, particularly those Nordics who had migrated to North America, and prove his theory. Unlike the dreamers and theorizers of the past, however, Henry Fairfield Osborn had the means to test his theory about Central Asia. Now all he had to do to complete the puzzle was to go there.

Notes

1. R.W.B. Lewis, *American Adam* (Chicago: University of Chicago Press, 1955).
2. Walt Whitman, 'The Song of Myself,' in *Leaves of Grass* (New York: 1855), 24.
3. Anne Rose, *American Victorians and the Civil War* (Cambridge: Cambridge University Press, 1992).
4. Robert Hughes, *American Visions: The Epic History of Art in America* (New York: Alfred A. Knopf, 1997), 137–205.
5. Thomas P. Slaughter, *The Natures of John and William Bartram* (New York: Alfred A. Knopf, 1996).
6. Peter Bowler, *Evolution: History of an Idea* (Berkeley: University of California Press, 1983).
7. Ronald Numbers, *Darwinism Comes to America* (Cambridge: Harvard University Press, 1998) and Cynthia Eagle Russett, *Darwin in America: The Intellectual Response 1865–1912* (San Francisco: W.H. Freeman & Co., 1976).
8. For the reception of Darwinism and the Metaphysical Club see Russett, *Darwin In America*, 47–78.
9. Edward Rafferty, *Apostle of Human Progress: The Life of Lester Frank Ward* (Ann Arbor, MI: UMI, doc. diss., 1999).
10. Edward J. Pfiefer, 'The Genesis of American Neo-Lamarckism,' *Isis*, 56: 184 (1965), 156–167.
11. Edward Drinker Cope, *Origin of the Fittest: Essays in Evolution* (New York: 1877), viii.
12. Edward Drinker Cope to his father, May 16, 1886, quoted in Persifor Frazer, 'The Life and Letters of Edward Drinker Cope,' *American Geologist*, 26:2 (August 1900), 67–128.
13. Peter Bowler, 'Edward Drinker Cope and the Changing Structure of Evolutionary Thought,' *Isis*, 68:242 (1977), 255.
14. Many historians have noted Cope's quirky, tongue-tying lexicon. The nomenclature made his already complex theory more obscure. Charles Darwin, after reading Cope and Hyatt's work several times and having Hyatt explain it to him, gave up in frustration. For Darwin's comments see Francis Darwin, ed., *The Life and Letters of Charles Darwin* (New York, Appleton's: 1903), Vol. 2, 233.
15. Bowler, 'Edward Drinker Cope,' 264.
16. HFO, 'Paleontological Evidence for the Transmission of Acquired Characters,' *Nature* (January 9, 1890), 228.

17. HFO, 'Nomenclature of Mammalian Molar Cusps,' *American Naturalist*, 22:262 (October 1888), 926–928. For Cope see Edward Drinker Cope, 'Origin of Specialized Teeth in the Carnivora,' *American Naturalist* (1879) and others in *The Origin of the Fittest*.
18. HFO, 'Organic Selection,' *Science*, n.s. 6:146, 585 (October 1897), 583–587.
19. HFO, 'Evolution of Mammalian Molars to and from the Tritubercular Type,' *American Naturalist*, 22:264 (December, 1888), 1067–1079.
20. HFO to William Berryman Scott, 7/16/1888, box V, OSNY.
21. HFO, *The Age of Mammals* (New York: Macmillan Co., 1910), 9.
22. HFO to Lucretia Osborn, 9/17/1888, box V, OSNY.
23. See correspondence for May and June 1887, box IV, OSNY.
24. HFO, 'The Hereditary Mechanism and the Search for the Unknown Factors of Evolution,' *American Naturalist* (May 1895), 418–439.
25. HFO, 'The Paleontological Evidence for the Transmission of Acquired Characters,' *Nature* (January 9, 1890), 227–228.
26. Ibid., 227.
27. HFO to Edward Poulton, 8/27/1891, box IV, OSNY.
28. HFO to Edward Poulton, 10/22/1888, box IV, OSNY.
29. HFO to Edward Poulton, 1/29/1889, box IV, OSNY.
30. HFO to Edward Poulton, 8/27/1889, box VI, OSNY.
31. HFO to Edward Poulton, 11/21/1889, box VI, OSNY.
32. HFO to Edward Poulton, 8/29/1889, box VI, OSNY.
33. HFO, 'Present Problems in Heredity,' *Atlantic Monthly*, 67 (March 1891), 353–364.
34. HFO, 'Organic Selection,' *Science*, n.s. (October 15, 1897), 583–587.
35. HFO, 'Search for the Unknown Factors in Evolution,' *American Naturalist* (May 1895), 418–439.
36. HFO, 'The Hereditary Mechanism,' 424.
37. HFO, 'Origin of the Mammalia,' *American Naturalist*, 32:377 (May 1898), 309–334.
38. C.L. Morgan, 'On Modification and Variation,' *Science*, n.s. 4:99 (November 20, 1896), 733–740. J.M. Baldwin, 'A New Factor in Evolution,' *American Naturalist*, 30:354 (June 1896), 441–533.
39. Rainger, 127–128. Bowler argues that organic selection was natural selection, but with intelligent guidance and personal initiative as driving factors.
40. HFO, 'Are Acquired Variations Inherited?' *American Naturalist*, 25:291 (March 1891), 191–216.
41. Michael Kennedy, 'Philanthropy and Science in New York City: the American Museum of Natural History, 1868–1968' (Ph.D. diss., Yale University, 1968). Also Douglas Sloan, 'Science in New York City, 1867–1907,' *Isis*, 71:256 (1980), 37.
42. HFO to Lucretia Osborn, 3/30/1893, box V, OSNY.
43. Garland Allen, *Life Sciences in the Twentieth Century* (New York: John Wiley & Sons Inc., 1975), 9.
44. Ibid., 18.
45. Ronald Rainger, Keith Benson and Jane Maienschien, *The American Development of Biology* (New Brunswick: Rutgers University Press, 1991), 3.
46. HFO, ' The Present Problem of Heredity,' 353.
47. Rainger, 135.
48. HFO, typed diary for 1903, box XXIV, OSNY.
49. HFO to T.H. Morgan, 3/8/1916, Henry Fairfield Osborn Papers, Correspond-

ence collection, Central Archive, courtesy of the American Museum of Natural History. Hereafter CAAM.

50. Garland Allen, 'T.H. Morgan and the Emergence of the New American Biology,' *Quarterly Review of Biology*, 44 (June 1969), 179.
51. T.H. Morgan, *Critique of the Theory of Evolution* (Princeton: Princeton University Press, 1916), 25.
52. HFO to Madison Grant, 4/24/1914, CAAM.
53. HFO to Lucretia Osborn, 2/22/1894, box VI, OSNY.
54. Virginia Osborn, 3/4/1894, box VI, OSNY.
55. Condolence folder, Box VI, OSNY.
56. HFO to Lucretia Osborn, 4/25/1894, box VI, OSNY.
57. Virginia Sturgis Osborn to HFO, 8/1894, box VI, OSNY.
58. For the baby's short life see letters in box VI, OSNY.
59. For HFO's views on education see 'How to Teach Evolution in Schools,' *School and Society*, 23:25–31 (1926); 'The Seven Factors of Education,' *Educational Review*, 32:56–82 (1906); and others.
60. *Atlantic Monthly*, lxvii:401 (March 1891), 353–361.
61. For Knight's life and art see Sylvia Czerkas and Donald Glut, *Dinosaurs, Mammoths, and Cavemen: the Art of Charles R. Knight* (New York: E.P. Dutton, 1982).
62. For early representations of extinct animal life see Martin Rudwick, *Scenes From Deep Time* (Chicago: University of Chicago Press, 1992).
63. HFO, *Cope: Master Naturalist* (New York: Princeton University Press, 1931).
64. Ibid., 385.
65. Ibid., 397.
66. For the quote see HFO to Lucretia Osborn, 4/2/1897, box VI, OSNY. For the reference to 'cystitus' see *Cope: Master Naturalist*, 461.
67. Ibid., 468.
68. HFO to William Berryman Scott, 4/11/1897, box VI, OSNY.
69. HFO, *Cope: Master Naturalist*, 587–589.
70. HFO, '"Mutations" of Waagan and "Mutations" of De Vries or [and] "Rectigradations" of Osborn,' *Science*, n.s. 23:844 (March 3, 1911), 328.
71. HFO to William Berryman Scott, 5/25/1903, box VII, OSNY.
72. William Berryman Scott, 'On Variations and Mutations,' *American Journal of Science*, 3rd edn, 48:355–374 (1894).
73. HFO, 'Orthogenesis as Observed from Paleontological Evidence Beginning in the Year 1889,' *American Naturalist*, 56:643 (1922), 134–143.
74. HFO to R.W. Gilder, 1/18/1903, box 72, Century Company Records, Olcott Papers, Manuscripts and Archives Division, The New York Public Library Astor, Lennox and Tilden Foundations. Hereafter Century.
75. HFO, 'Evolution as it Appears to the Paleontologist,' *Science*, 26:744–749, 746 (1907).
76. HFO to Lucretia Osborn, 4/17/1906, box VIII, OSNY.
77. HFO, 'Coincident Evolution Through Rectigradations and Fluctuations,' *Science*, 27 (1908), 749–752.
78. HFO to T.H. Morgan, 5/27/1910, CAAM.
79. HFO to T.H. Morgan, 6/11/1918, CAAM.
80. HFO to Richard Gilder, n.d. [1905?], Century.
81. HFO, *The Origin and Evolution of Life* (New York: Scribner's Sons, 1917), vii.
82. HFO to William Beebe, 8/2/1920, folder 5, box 16, William Beebe Papers, PU.

83. HFO to William Beebe, 11/29/1920, folder 5, box 16, William Beebe Papers, PU.
84. See letters between HFO and Henry Fairfield Osborn Jr, box VII, OSNY.
85. HFO to Lucretia Osborn, 1/16/1906, box VIII, OSNY.
86. HFO to Lucretia Osborn, 4/18/1905 , box VIII, OSNY.
87. HFO to Josephine Osborn, 2/14/1908, box VIII, OSNY.
88. HFO to Lucretia Osborn, 4/18/1901, box VII, OSNY.
89. HFO to Lucretia Osborn, 6/6/1905, box VII, OSNY.
90. George Gaylord Simpson, *Concession to the Improbable: An Unconventional Autobiography* (New Haven: Yale University Press, 1978), 40.
91. HFO to Lucretia Osborn, 8/14/1906, box VIII, OSNY.
92. See letters from HFO to Josephine Osborn, December, 1906–February, 1907, box VIII, OSNY.
93. Irvin Wyllie, *The Self-Made Man in America: the Myth of Rags to Riches* (New Brunswick: Rutgers University Press, 1954).
94. HFO, 'Ontogenetic and Phylogenetic Variation,' *Science*, n.s., 4:100 (November 1897), 789.
95. Allen, 'T.H. Morgan,' 26.
96. James Moore, *Post-Darwinian Controversies* (New York: Cambridge University Press, 1979), 270–276.
97. For the life of Gray see A. Hunter Dupree, *Asa Gray: American Botanist, Friend of Darwin* (Baltimore: Johns Hopkins University Press, 1988).
98. Moore, *Post-Darwinian Controversies*, 334.
99. HFO, 'The Four Inseparable Factors of Evolution,' *Science*, n.s., 27:682 (January 24, 1908), 148–150.
100. HFO, 'Evolution as it Appears,' 744–752.
101. Bowler, 'Edward Drinker Cope,' 175.
102. HFO, 'Coincident Evolution,' 749–752.

CHAPTER FOUR

The Central Asia Hypothesis

By the early 1900s Henry Fairfield Osborn felt he had come very close to proving his general theory of evolution, though few other scientists concurred. They were moving off in other directions, led by the new biology, genetics and neo-Darwinism. He saw that the new biology with its distaste for morphology and fossils and its reliance upon laboratory techniques and the study of heredity at the cellular level, was proving a more insightful way of approaching the subject and was proving his theories obsolete. He knew he was being abandoned by the scientific mainstream, even by some of his closest allies. His assistant, William Diller Matthew, being more of a traditional Darwinist who was not interested in internal perfecting mechanisms, was drifting away. Osborn lamented to Wick Scott: 'I find my conclusions so different from those of Dr Matthew, for example, that … all of his conceptions are quite different from mine.'[1] The new ways, with their loss of certainty and abandonment of old authorities, frightened him, while the old ways still comforted him. In the end he danced around the scary parts, leaving them to younger more fearless researchers, and stuck with what eased his mind and soothed his soul.

Undaunted by the situation, Osborn struggled on. He had to make the crucial link between mammal and human evolution. In order to complete the formula he had to find a center where the first humans had appeared, and from whence they could overcome hardship and achieve evolutionary advance. As time would go by that center looked more to him like Central Asia. If he could prove where humans had first appeared, the rest of the puzzle would fall into place. If he could not, the entire façade would topple like a house with no foundation.

During the summer of 1901 Osborn's daughter Josephine was staying with Mother Osborn. A high-spirited young lady, Josephine was anxious to get back to traveling and told her mother, 'I am glad to think this is my last summer with Grandma.'[2] It was a prophetic statement. On January 26, 1902 Virginia Osborn suffered a heart attack. On February 7, she died quietly.[3] Though her friends reacted with great sorrow and made emotional speeches and eulogies at her passing, there was little outward reaction from her eldest son. With middle-class, Yankee restraint he recorded a brief, matter-of-fact notice of her death in his diary. As he did with the deaths of his father and siblings, Henry Fairfield Osborn kept his feelings private, and chose to obliterate his pain through more hard work.

Osborn had two important assistants in William Diller Matthew and William K. Gregory. Matthew became Osborn's assistant in the 1890s. He made crucial contributions to the study of faunal distribution and migration. Matthew saw these areas as a major ingredient in understanding continuous evolutionary processes. Like Osborn, he believed that the more simplified forms of organisms were closer to the group's point of initial radiation. His work cleared up several questions of biostratigraphy that had plagued Osborn.[4] His distribution work, and his belief in the Central Asian origins of mammals, greatly assisted Osborn's theory, but he did not agree with Osborn on everything. Matthew saw climate rather than some internal mystical process as a central engine of evolutionary change. He did not think evolution predetermined.[5] William K. Gregory also followed Osborn's suggestions about what to study, but, like Matthew, eventually parted company with the professor over key issues of evolution. Gregory first became interested in human evolution when Osborn asked him to examine and write a paper on the 1912 Bristish discovery, Piltdown Man.[6] Gregory (1876–1970) had originally planned on becoming an Episcopalian minister. Instead he attended Columbia College and became interested in natural history after taking classes on the subject, including Osborn's. Impressed, Osborn made Gregory his assistant in 1899. He became a full museum science staff member in 1911, and then an instructor at Columbia in 1916.[7] Both Matthew and Gregory helped Osborn extensively and contributed both to his research and to his writing. He could use all the help he could get.

Despite the emotional upheavals in his life Osborn was making what he thought was headway in his evolution program. Working with William Diller Matthew on fossils from the Wasatch and Puerco formations of North America Osborn had 'reached the most important results, which are absolutely revolutionary in character' about the ancestry of mammal faunas in these localities.[8] His explanation for the process whereby early generalized forms evolved into a series of more specialized varieties and filled an otherwise open geographical area he dubbed 'Adaptive Radiation.' It was an idea he began developing in the early 1890s. It held that a small unspecialized type of mammal could enter a geographic area and then, if conditions were right, spread out, or radiate, to fill the region with more specialized variants.[9] Once established in a region, a group would then begin to show its evolutionary trends. Adaptive Radiation was powered by his 'Law of Four Inseparable Factors' of evolution and contained the important element of animal migration.[10] Organisms, he thought, had to spread out across the landscape in order to prosper.

An obstacle to worldwide migration theories was the fact that though fossils of the same types of animals had been found on different continents, there seemed no way for them to have moved from one land mass to another. A way around this dilemma was to postulate the existence of land bridges – connections between continents that rose and fell at different periods. Since the 1830s geologists such as Edward Forbes, Charles Lyell and others saw the changing landscape and coastlines of the world as a result of the vertical rise and fall of land and sea. Periodically, they believed, the waters would rise, inundating parts of the earth cutting off one region from another. When the waters receded land masses were reconnected, allowing for the migrations of animals and people. When the cycle was repeated it gave the illusion that either similar species had arisen independently on different continents or that the continents had been connected at one time. Since the oceans did rise and fall, and floods did occur, it seemed a reasonable explanation for the baffling problem of faunal distribution. Land bridges proved useful tools: if one could not figure out how an animal group or group of people or even an entire culture got from one continent to another, one could simply fall back on a now sunken land bridge. If land bridges could rise and fall, why not lost continents like Atlantis or Lemuria?[11] Ernst Haeckel thought Lemuria the birthplace of the early primates (as did Madame Blavatsky). Henry Fairfield Osborn especially liked land bridges because they suggested slow and steady rather than abrupt changes in the earth's structure. Land bridges allowed for animal migrations with 'no very profound or cataclysmic changes … to connect the Northern continents with each other and with outlying masses.'[12] He was happiest with the idea of a conservative and methodical universe.

The year 1900 saw the publication of one of Osborn's pivotal papers. In 'The Geological and Faunal Relations of Europe and America During the Tertiary Period and the Theory of the Successive Invasions of an African Fauna,' he suggested the Central Asia hypothesis.[13] Building on the work of W.D. Matthew, Osborn posited that a land bridge must have existed in the past that linked South Africa, South America, Australia and New Zealand to Indonesia, which allowed organisms to migrate from Asia to the New World. Citing the work of others such as Forbes, Milne-Edwards and Florentino Amagheno, he considered the existence of a Holarctic, or northern, land bridge to be a proven fact. The assumption that land bridges had connected the continents provided the basis for his argument about where various groups had originated. He proposed that the more specialized the variant, the further it was from its point of origin. The more general forms had to be the oldest and therefore closest to the source of original radiation. Following this logic, and the state of the fossil record, he suggested that looking to Asia would lead to answers about the origins of mammals. Though he would later claim otherwise, the 1900 paper is only a general statement, and does not contain great detail. In 1925 he called it his 'paleontological prophecy' although he never explicitly said Central Asia was the home of mammals or humans. Since man was a mammal, the first humans would also have come from Central Asia. He did say that a thorough exploration of this area would yield important results in the search for their home.

As would be expected, one of the foundations of his Central Asia hypothesis was the story of the horses. He studied their fossil remains most of his adult life and saw them as key to understanding the course of mammalian evolution (he loved riding horses too, though he had a nasty habit of falling off them at inappropriate moments). The fossil beds of the Western United States were a treasure trove of horse fossils. Complete fossil series of many different types of horses were discovered which seemed to prove that evolution was a slow, steady, almost predictable process of patterns and trends over eons of time. They seemed to offer irrefutable proof not only of the fact of evolution, but for its sequence as well. Osborn also spent years studying the titanotheres, which he saw as even more integral to the mammal evolution story. Huge, lumbering, rhinoceros-like creatures, these 'Thunder Mammals' were closely related to the horses, rhinos and tapirs, and had inhabited North America and parts of Asia. He was particularly fascinated by the titanothere's unusual horn development, which ranged from stubs to elaborate forms as big as the animal's head. He saw this horn-size progression as support for his theory. He surmised that the phylogenetic change in horn size and shape was also correlated with the changing adaptive function of the horns. When the horns were small, as in the early forms, they did not affect the creature's life. As time went by, the horns grew larger and thus became a benefit for use in mating and fighting. As evolution continued its inevitable trend toward a larger and larger size, the horns adversely affected the phyla to a point where they were no longer a benefit, but a hindrance. They became so large, he argued, that they drove the group to extinction because their overspecialization made them unable to adapt to their environment.

Osborn worked on the titanotheres over the course of his career, an effort that resulted in his magnum opus, *The Titanotheres of Ancient Wyoming, Dakota, and Nebraska* (1929). The two-volume work – itself titanic – took decades of research and writing, and tens of thousands of uncle Morgan's dollars to complete. It was the largest writing project of his life (along with an equally ponderous two-volume work on extinct elephants published at the end of his life which reportedly cost Morgan almost $300,000.00). It was always on his desk and on his mind, in one form or another. Both Matthew and Gregory worked on it as well as other museum staff. He felt that explaining the evolution of the titanotheres would explain the evolution of all mammals and even humans, but he was having trouble working out that explanation. 'The work is very difficult,' he confided to his wife, 'and I have not yet succeeded in establishing a theory which will explain all the facts.'[14] He traveled to Colorado to examine specimens there, which brightened his spirits. He told Loulu that the trip was 'absolutely essential, both for my knowledge and reputation,' and that it spurred him to greater effort. The book went slowly in part because of his endless administrative duties as president of the museum. Because of these distractions, he would often draw a blank. He knew that if he kept at it, though, inspiration would 'burst upon me.' In the spirit of Darwin he accumulated facts and waited for 'an induction' to come along.[15]

He had other forms of assistance besides the museum staff. Once, while working through a particularly difficult section, he wrote Loulu. 'You can help me by praying for it and coming down [to see me].' He told her, 'I want you to break all engagements and come down by the noon train Saturday' and we can have 'a little wedding trip' before returning to work.[16] Thus refreshed, he felt he had broken through and his ideas about the ancient beasts were 'proving true.'[17] When it finally did appear, Osborn called the publication of the titanothere volumes 'the greatest event of my life.'[18]

The titanotheres were central to Osborn's evolutionary thinking because in their structure and behavior he saw tangible proof of his ideas. The development of their teeth and horns showed, he said, how the titanotheres had grown from simple unspecialized forms, which adapted easily to their environment, to overspecialized monstrosities whose complex tooth and horn structure condemned them to extinction because they were no longer fit (this was the same concept he would later apply to human development). They also seemed to follow the concept of the irreversibility of evolution put forward by Belgian paleontologist Louis Dollo (1857–1931) and adopted by Osborn. Osborn's interpretation of Dollo's theory said that once a character began to evolve in a certain direction (horns getting larger for example), they could never go in a different direction (such as getting smaller). Though this was not quite what Dollo meant, that is how Osborn interpreted it.[19] Also, titanothere migration patterns seemed to follow his law of adaptive radiation. They seemed tailor-made to prove all of Osborn's evolution theories.

That mammals had originated in Asia and then migrated around the world was likewise supported by spectacular fossil finds being made in South America, particularly Patagonia. On his 1830s voyage, Charles Darwin had collected some of the first known South American fossils. The continent was the home to strange creatures like the giant ground sloth and the glyptodon – a sort of armored teddybear. Osborn postulated that such specialized creatures must be far from the mammal's original home. He built upon the pioneering work of Argentine paleontologist Florentino Ameghino (1854–1922), who discovered and described much of South America's extinct fauna. Osborn sent expeditions to Patagonia, as did William Berryman Scott of Princeton. For Osborn, Patagonian fossils showed that there had been a connection of northern continents to the huge southern continent of Antarctica. This southern landmass, still a hypothesis as late as the 1890s, connected the Americas to Australia, he said. At the top of the world, Osborn argued, North America was connected to Asia and Europe via the huge Holarctica (the North Pole). He considered this massive area to be the 'Grand Central [Station] of evolution,' the highway used by mammals, including humans, to get from their point of origin in Central Asia to the rest of the world.[20]

Though a respected paleontologist, Ameghino had developed an interesting, if outlandish, theory of human-origins. Like the works of many human-origin theorists discussed in this study, Florentino Ameghino's theory had nationalist and patriotic overtones as well as a scientific basis. He argued that the first humans had come from Argentina. He based his theory on controversial evidence of early human habitation in South America dated to the Pliocene and Miocene.[21] His work attracted worldwide attention and several scientists traveled to Argentina to examine that evidence. Smithsonian Institution ethnologist Ales Hrdlicka (1887–1943), the bane of human-origin theorists, dismissed Ameghino's work.[22] A Bohemian-born physician, Hrdlicka had turned to anthropology and written a book in 1907 carefully dissecting claims of fossil humans, most of which he found lacking foundation.[23] Hrdlicka was not alone in his reaction to Ameghino. William Berryman Scott, who had done extensive studies of Patagonian fossils, was also skeptical. Though he too respected Ameghino's other work, he found his human-origin theory 'an utterly inconceivable hypothesis.'[24] Osborn, so wedded to his Central Asia concept, likewise saw little merit in it and continued on his own path. Along the way Osborn bumped into an unlikely traveler.

Nebraska Man

In the early years of the twentieth century, Osborn had begun quietly to formulate an idea about an early human ancestor he would eventually call 'Dawn Man,' that would cap his work on evolution.[25] Since this creature was a hypothetical construction, he needed some tangible evidence to give it credence. Osborn's first fossil human came from a familiar place. In 1906, Richard Gilder, a Nebraska journalist and amateur fossil hunter, was searching for human fossils near Indian burial mounds outside of Omaha when he discovered bones with unusual characteristics. He took them to Nebraska state geologist Edwin Barbour along with other remains found by an earlier group of mound busters in 1894. An article about the discoveries then appeared in the *Omaha World Herald*, a copy of which made its way to Osborn through Gilder's New York relatives. Osborn was so excited by what he read that he immediately dropped everything and went to Nebraska to view the material first hand.[26] Once in Nebraska, he held up the skullcap and looked deeply into its empty half-eye sockets and thought he saw eternity.

Osborn had not done much direct work on fossil humans previously. His research, begun in 1900 or so, was more theoretical than practical. He had followed the work of Eugène Dubois, Marcellin Boule and other physical anthropologists, but he had little hands-on human fossil experience himself. What added to his excitement was that the Nebraska find came from a locality he was already familiar with through his study of fossil horses. Osborn suspected that if environmental conditions were right, humans could have lived there in the past, as he saw horses and primates following similar

322 NEBRASKA GEOLOGICAL SURVEY

Fig. 2. Skull No. 6. Nebraska Man, front view, showing thick protruding superciliary ridges, frontal bone without eminences, right and left parietals. Fronto-parietal and interparietal sutures deeply dentate. Three-fifths natural size.

Specimen No. 6-1-11-06. The Robert F. Gilder collection, State Museum.

Negative No. 10.5-1-11-06. Hon. Charles H. Morrill's collection of geological photographs.

Fig. 5 The first Nebraska Man. Osborn's initial excitement over this find quickly turned to disappointment.

Photo from the Nebraska Geological Survey publication which heralded it.

migration patterns. Similar conditions in Europe had produced what was thought to be evidence of human habitation, so why, he speculated, could the Americas not have done the same?

Osborn met with Gilder, Barbour and Henry Ward at the University of Nebraska where the bones were deposited. The unusual characteristics – heavy brow ridges on the skullcap and general robustness of the bones – suggested their great antiquity. With Barbour and Ward preparing a detailed scientific description for publication, Osborn went to the public forum of *Century Magazine*. Barbour and Ward's technical description appeared in December 1906. They said the find, which they referred to as Nebraska Man, might be 'the earliest type of man known as yet in America.'[27] Osborn's *Century* article appeared the following month. In it he said that anthropologists were of two minds about the appearance of humans in the Americas. They believed they had either appeared very early or very late. He placed himself in the former camp. Since he believed a land bridge connecting the Americas to Asia and Europe had allowed for the migration of animals during the Pleistocene age, he concluded that humans must have done the same. He argued that there were 'different races of men' all vying for survival in a hostile environment, and some undoubtedly made it to the New World. Osborn felt that Nebraska Man was more advanced than the Neanderthals, who he claimed were of a different species and not direct ancestors to modern humans. He also argued that there were tantalizing scraps of evidence (in the form of animal bones and geological features) which, while not convincing in themselves, suggested the possibility of early human habitation in North America. He lamented that the other 'men' – like Lansing Man and Calvaras Man – had been so disappointing. Nebraska Man, however, seemed more promising.[28]

Simultaneously with Osborn's *Century* article, the other Nebraska men – Gilder, Barbour and Ward – published their own articles in *Putnam's Magazine*. They described how the bones were found, their physical attributes, and even something of the creature's lifestyle. Barbour said Nebraska Man must have been 'low and savage,' but advanced enough to have religion and veneration of the dead.[29] All involved were convinced they had something very special on their hands.

Like an ill wind, however, wherever claims of ancient humans appeared Ales Hrdlicka was soon to show up. When he turned his critical eye on them, few failed to wither under the gaze. Sent by his boss at the Smithsonian to view the bones, Hrdlicka examined all the Nebraska Man material with an eye towards anatomical detail. After a lengthy description, Hrdlicka got to the bad news. 'If the present knowledge concerning these specimens is impartially considered,' he said, 'it is apparent that the theory of a more than recent geological origin of any of them meets with serious objections.'[30] In addition to the bones being mundanely like any other Indian burial, Hrdlicka pointed out what had apparently not been obvious to the discoverers: the bones had not fossilized. Stymied, Osborn could only accept Hrdlicka's reasoned assessment, forget Nebraska Man, and hope for better news to come along.

Towards the conclusion of his 1910 book, *The Age of Mammals* – his major comprehensive statement of his views up to that point – Osborn considered the presence of early humans in the Americas. Tantalizing finds thought to support the antiquity of early man had been appearing since the 1830s though none were conclusive. The nineteenth century was a heady time for discovering fossil humans, at least what people thought were fossil humans. A parade of 'Men' appeared in North America: Lansing Man in California, Vero Man in Florida, and Trenton Man in New Jersey, among others. The English had Galley Hill Man while the South Americans had Ameghino's Tetraprothomo and Diprothomo. They all turned out to be misidentifications of modern skeletons. Hrdlicka was convinced there was no credible evidence to support the belief that early humans had inhabited the New World. Osborn admitted that Hrdlicka was right, but refused to give up hope. 'This does not mean,' Osborn said, 'that early man did not exist in North America [only] that convincing proof … still remains to be produced.'[31] Even though Hrdlicka had quashed any thought of early humans in the Americas, Osborn was only just getting started in the quest for his ancient ancestors.

Fossil Men

Genuine human fossils were first discovered in 1865 when a primitive skeleton was uncovered in the Neander valley of Germany. Neanderthal Man was a controversial enigma, and there was little consensus as to its identity. In 1868 more human-like fossil remains were unearthed in Les Eyzies, France. They were more modern looking than the Neanderthal, yet still seemed archaic. Dubbed Cro-Magnon Man, the creature fell somewhere between modern humans and the Neanderthal. Cave paintings and artifacts belonging to the Cro-Magnons were also discovered. In 1886 more Neanderthal-like remains were found near Spy, Belgium. This proved that the original find had not been a freak of nature as some thought, but representative of a group. Whatever they were, the Neanderthals and Cro-Magnons were discovered in Europe and as such did not seem to shed any light upon the Central Asian origins of humans.

That all changed in 1891. German naturalist Ernst Haeckel lectured widely in the 1870s that man had descended from an ape-like ancestor. Since there were few Neanderthal or Cro-Magnon fossils known at the time (and they were relatively young anyway), Haeckel postulated a hypothetical human ancestor, closer to the primates on the evolutionary line, to fill the gap. His work inspired Dutchman Eugène Dubois (1858–1940) to search for the remains of this creature. Though Charles Darwin, in *The Descent of Man* (1871), suggested that Africa was the place to look for the earliest humans, Haeckel said that since humans were related to the apes and apes live in the tropics, Asia was where the search should concentrate. Dubois, a professor of anatomy in Amsterdam, set off to Sumatra to find the missing link in 1887.

After several fruitless years working a site along the Solo River near the village of Trinil, Dubois found what he was looking for in 1891. Consisting of a skullcap, femur and a tooth (the same inventory as Nebraska Man), the find seemed to be a man-like ape much more primitive than the Neanderthals. Using Haeckel's term, the find was labeled *Pithecanthropus*, but was commonly called Java Man. Dubois insisted that Java Man was the missing link between humans and apes. Haeckel claimed that Dubois's find proved that humans had descended from the primates. Not everyone agreed, however.

A debate began over how to interpret Java Man. Problems arose with the fossils themselves. Though he tried to work systematically, Dubois did not keep careful records of the strata in which the fossils were found. Also, the fossils had not all been found in the same spot, making it impossible to be sure they belonged to the same skeleton. Dubois frantically showed his fossils to everyone he could, trying to convince other scientists that his discovery was genuine. Nevertheless Java Man did support – at least superficially – the Asian origins of humanity. It was, in fact, the only remotely plausible bit of evidence uncovered so far to do so.[32]

Additional finds began to appear. In 1912 amateur naturalist Charles Dawson claimed to have discovered human fossils in the Piltdown gravel quarry in Sussex, England. Several skull and jaw fragments along with eoliths (stone tools) were found. Dawson showed the material first to Jesuit anthropologist Pierre Teilhard de Chardin and then to anatomist Arthur Smith-Woodward, who labeled the find *Eoanthropus dawsonii* (Dawson's Dawn Man). Typical of every new find, a debate ensued over the discovery: first over the dubious and rumor-shrouded discovery details, and more importantly over the curious fact that the skull seemed very human while the jaw was much more simian. Some British scientists were quick to support Piltdown Man in part because it seemed to contain all the elements necessary for a missing link, and partly because of nationalist feelings (they were proud that such a creature was British). Although the fossils were an anomaly that did not fit properly into the human family tree, doubters were lost in the clamor of praise for the discovery. Piltdown Man was eventually found out to be a hoax in 1950, but during Osborn's lifetime it was generally accepted as genuine.[33]

Despite their initial skepticism, by 1921 both Osborn and William K. Gregory accepted Piltdown as genuine, based upon their inspection of the original pieces in England. Smith-Woodward sent Osborn casts of the Piltdown material as well. Osborn became an avid collector of fossil human casts and used them in his work. While Smith-Woodward happily gave the copies, French anthropologist Marcellin Boule refused to give copies of his Neanderthals.[34] Eugene Dubois also refused at first, but after several years of bullying and conniving by Osborn, relented.[35] Osborn only became interested in Ales Hrdlicka's efforts to raise money to help the scientific community in his home region of Bohemia after World War I, when he thought Hrdlicka might be able to procure casts of German fossil humans in return.[36] Human ancestry was increasingly at the forefront of Osborn's work.

Men of the Old Stone Age

Once Osborn had developed a series of scientific ideas to explain mammal evolution, it was time to apply those ideas to human evolutionary questions. His early work on paleoanthropology culminated in *Men of the Old Stone Age* (1915).[37] The book was meant to be the foundation of his human evolution theory, yet contains little of his own original work. It was primarily a survey of the latest paleoanthropology done by European researchers, particularly the Frenchmen Marcellin Boule and Henri Breuil.[38] The two Frenchmen had done extensive work on archaic human life of France. Boule worked on Neanderthals and Breuil was the leading authority on the cave art and artifacts of Cro-Magnon culture.[39] Breuil worked with Osborn regularly and gave Charles R. Knight pointers on the illustrations for the French edition of the book.[40] William D. Matthew and William K. Gregory also went over the original manuscript. They were uneasy with some of Osborn's assertions and quietly told him so. Impressed with their critiques, Osborn wanted to include them as an addendum. Neither man would go along with the request as they thought it bad form to criticize their boss publicly.[41] Years later, however, Gregory in particular grew more willing to joust with Osborn in public.

Osborn looked to fossils as well as to cave art in building his case for how humanity and its society spread across Asia and Europe. *Men of the Old Stone Age* is less a study of the origins of humans than a discussion of how, once having appeared, they grew as sentient beings. Though Osborn focused on early people in Europe, he stated flatly that Asia was 'the chief theater of evolution of both animal and human life.'[42] The book proved a popular success (his personal best-selling work) and went through several reprints and re-issues. People brought it with them when they toured the museum; author Edith Wharton carried it with her as a guide when she toured the Cro-Magnon sites of France; and H.G. Wells used it as a source work for his *Outline of History*.[43]

The structure of *Men of the Old Stone Age* was to present levels of evidence all carefully crafted to support Osborn's ideas about the evolution of humans. Building upon *The Age of Mammals*, he explained how mammals originated in Central Asia and then migrated around the world by using land bridges to get from one continent to another. This became his standard tactic: explain how animals evolved and spread, and then apply those rules to humans. He said that both ape and human ancestors had first appeared in Central Asia and then migrated west into Europe and Africa. He also addressed climatology, arguing that the climate of Asia worsened as the Pliocene age gave way to the Pleistocene. With this change from a temperate to a harsh environment, hardier species thrived while those less able to cope went into decline. The more advanced animals began to migrate west, opening the way for the later successive waves of early humans that followed. Osborn estimated that the first humans entered Europe between the third and fourth Ice Ages, about 70,000–100,000 years ago.

Conclusion

His study of the prevailing wisdom led Henry Fairfield Osborn to believe that various forms of fossil humans had first appeared in Central Asia, yet were not directly related to one another. He argued that the Neanderthals and Cro-Magnons had risen from different stock. The Cro-Magnons were a superior race whose advanced intellect allowed them not only to out-class the Neanderthals, but to create a complex social order with an appetite for beauty and the ability to create art. Their artistic ability was an outward sign of their inner superiority and advanced biology which, as in all organisms, was the result of their race plasm. The Cro-Magnons had a 'unity of heredity' shown by the 'unity of their art impulse,' which could not have been passed on to them from another race, but was inborn, the result of their predestined characters.[44] Osborn posited that a number of different human groups had appeared from different ancestral stock and then in a series of waves radiated out, displacing the group that had gone before.

In the end even the vaunted Cro-Magnons were bested by the final wave of humans who swept out of Central Asia to become the superior form of modern humans. 'The sudden appearance in Europe,' said Osborn, 'of a human race with a high order of brain power and ability was not a leap forward, but the effect of a long process of evolution elsewhere.' Echoing his 'prophecy' of 1900, he concluded that 'when the prehistoric archeology of Eastern Europe and of Asia has been investigated we may obtain some light on this antecedent development.'[45] Though he did not say so explicitly, Osborn was paving the way for the Aryans and Nordics he felt constituted this superior wave. These were the same Aryans beloved by the neo-Romantic French and German philosophers from Hegel to Schlegel to Gobineau, those Ernst Haeckel patterned himself after, and those T.H. Huxley had scoffed at. They became the hope for mankind's ascendancy for Madame Blavatsky, and the hope that salvation could be earned for Henry Fairfield Osborn. While Osborn's work on the origins of man had its core in science and the fossil record, other less sober ideas and evidence were being added.

Notes

1. HFO to William Berryman Scott, 2/5/1921, box IX, OSNY.
2. Josephine Osborn to Lucretia Osborn, 8/25/1901, box VII, OSNY.
3. Typescript copy of HFO diary for 1902, OSNY.
4. W.D. Matthew, 'A Provisional Classification of the Freshwater Tertiary of the West,' *Bulletin of the American Museum of Natural History*, 12 (1899), 19–75, and 'New Canidae from the Miocene of Colorado,' *Bulletin of the American Museum of Natural History*, 16 (1902), 286–287. Also see Ronald Rainger, 'Just Before Simpson: William Diller Matthew's Understanding of Evolution,' *Proceedings of the American Philosophical Society*, 130 (1986).
5. W.D. Matthew, 'Hypothetical Outlines of the Continents in Tertiary Times,' *Bulletin of the American Museum of Natural History*, 22 (1906), 353–383.
6. William K. Gregory, 'The Dawn Man of Piltdown, England,' *American Museum Journal*, 14 (1914), 189–200.
7. Edwin Colbert, 'Biographical Memoir of William K. Gregory,' *National Academy of Science*, 46 (1975), 91–133.
8. HFO to William Berryman Scott, 2/20/1897, box VI, OSNY.
9. HFO, 'The Rise of Mammalia in North America,' *Nature*, 48:1246 (September 14, 1893), 462, and HFO, *Age of Mammals*.
10. HFO, 'The Law of Adaptive Radiation,' *American Naturalist*, 36:425 (May 1902), 353–363.
11. S.L. De Camp, *Lost Continents: The Atlantis Theme in History, Science, and Literature* (New York: Gnome Press, 1954).
12. HFO, *The Age of Mammals*, 20.
13. HFO, 'The Geological and Faunal Relations of Europe and America During the Tertiary Period and the Theory of the Successive Invasions of an African Fauna,' *Science*, n.s., 11:276 (April 13, 1900), 561–574.
14. HFO to Lucretia Osborn, 4/11/1901, box VII, OSNY.
15. HFO to Lucretia Osborn, 5/24/1901, box VII, OSNY.
16. HFO to Lucretia Osborn, 5/11/1901, box VII, OSNY.
17. HFO to Lucretia Osborn, 1/25/1902, box VII, OSNY.
18. HFO to Charles Scribner, 12/23/1929, Scribner Archive, folder II, box 108, PU.
19. Dollo argued that once an organism lost a character or organ through evolution, it could never be regained.
20. HFO, *Age of Mammals*, 65.
21. Florentino Ameghino, 'L'age des Formations Sedimentaires de Patagonie,' *Anales de la Sociedad Cientific Argentina*, 50 (1900), 109–130, 145–165, 209–229; 51 (1901).
22. Ales Hrdlicka, *Early Man in South America*, Bureau of American Ethnology Bulletin, 52 (Washington: Government Printing Office, 1912).
23. Ales Hrdlicka, *Skeletal Remains Suggesting or Attributed to Early Man in North America*, Bureau of American Ethnology Bulletin, 33 (Washington: Government Printing Office, 1907).
24. Scott, *Memories of a Paleontologist*, 251.
25. HFO, 'On the Origin and Contemporary Evolution of Man,' *Record Herald of Chicago* (1902), and 'The Cartwright Lectures for 1892 before the Alumni of the College of Physicians and Surgeons, New York,' *American Naturalist*, 26 (1892), 455–481.
26. HFO to William Beebe, 10/25/1906, William Beebe Papers, PU.

27. Erwin Hinckley Barbour and Henry Baldwin Ward, 'Preliminary Report on the Primitive Man of Nebraska,' Nebraska Geological Survey (December 22, 1906).
28. HFO, 'Discovery of a Supposed Primitive Race of Men in Nebraska,' *Century Magazine* (January 1907), 371–375.
29. *Putnam's Monthly* (January 1907), 407–415, 502–503.
30. Hrdlicka, *Skeletal Remains*, 66–98, quote on 87.
31. HFO, *Age of Mammals*, 499.
32. For Dubois see Pat Shipman, *The Man Who Found the Missing Link* (New York: Simon and Schuster, 2001).
33. Charles Blinderman, *The Piltdown Inquest* (Buffalo, NY: Prometheus Books, 1986).
34. HFO to Arthur Smith-Woodward, 1/12/1915, CAAM.
35. Shipman, *Man Who Found the Missing Link*, 354–366.
36. See correspondence between HFO and Hrdlicka, 6/1920, CAAM.
37. HFO, *Men of the Old Stone Age* (New York: Charles Scribner's Sons, 1915).
38. It was also the result of many staff members' work. A large portion of all of Osborn's published books and articles were written by others, at least in outline form if not word for word. He often, though not always, gave credit to those workers in the small type.
39. Alan Houghton Broderick, *The Abbé Breuil, Prehistorian* (London: Hutchinson, 1963).
40. Henri Breuil to HFO, 1/27/1919, Charles R. Knight Papers, box 1, Manuscripts and Archives Division, New York Public Library, Astor, Lenox and Tilden Foundations. Hereafter NYPL.
41. See correspondence between Osborn, Gregory and Matthew, 12/1914, CAAM.
42. HFO, *Men of the Old Stone Age*, 19.
43. For sales figures see Scribner Archive, author files, box 108, folders 3 & 4, PU. He debated a couple of different titles, including *The Dawn of Man*, *Early Man in Europe*, and *Early Man: His Art and His Tools*. HFO to Charles Scribner, 11/7/1913, PU. Edith Wharton to HFO, 6/29/1923, box IX, OSNY.
44. HFO, *Men of the Old Stone Age*, 315.
45. Ibid.

CHAPTER FIVE

A Mongrelized World

For her first communion in May of 1908, Henry Fairfield Osborn threw his daughter Josephine an elaborate party and invited the entire neighborhood, including 'people of all classes.'[1] It is interesting that he would have congratulated himself on having done so. Osborn has long been associated with anti-immigrant feelings and eugenics legislation. He certainly held racist and anti-Semitic views. The question is why and to what extent. Beginning in the 1890s, increased immigration, a perceived societal breakdown in morality, and the advent of the modern age changed the American landscape in ways that disturbed him. In his mind the human species was being ruined by the same things – overspecialization, miscegenation and a loss of creativity – which he believed had doomed other species to racial collapse and extinction.

It has been argued that Osborn's views on evolution were colored by, if not prompted by, his racism. The situation is more complex. His racial beliefs were a product of a complex interaction of his own particular evolution theory with other ideas layered over the top of it and then mixed in a symbiotic exchange. Racism worked its way into his racialism as he and racist eugenicists and immigration restrictionists swapped ideas. As time progressed, Osborn's views evolved from a benign paternalism to a harder attitude and then to a increasing appreciation of human cultural and social diversity. His work on the horses and titanotheres showed him that organisms which seemed to be very much alike were usually different species with different abilities to adapt to and overcome their environments. Osborn's general theory of mammal evolution was complete before he began applying it to humans. With a belief that human groups were different species, it was easy for him to accept the principles of eugenics and immigration restriction. At its core, Osborn's path was laid down by his scientific studies, not by race hatred. Yet on this issue, Osborn is not an easy fit. He could be dismissive of other ethnic groups, yet then be supportive of them. In a speech given during the early days of World War I he addressed the negative influence of prejudice and hatred on society, and how the teaching and study of the past could be used to combat it. 'History must be taught,' he said, 'not to perpetuate hate and prejudice, but to eliminate these pernicious influences from our national life, to develop clear thinking, to vindicate right, and promote justice.'[2] After the war, however, he remarked that the conflict had shown that 'the Negro is not like us' and that 'we learned that some [ethnic groups] we had believed possessed an order of intelligence perhaps superior to ours were far inferior.'[3] He approached the changes in society in the same way as he did those in nature. America, and in particular his home of New York City, was not following nature's rules, and was progressing far too fast for him to be comfortable with.

New York City

Osborn's evolution theory was a combination of blood aristocracy and the essential American characteristic of meritocracy. Accepting the existence of predetermined biological advantages, Henry Fairfield Osborn also believed in the efficacy of personal struggle. He lived most of his life in the city that epitomized change and the benefits of that struggle. In New York people could take whatever they had been born with and build themselves into something better. In the early decades of the twentieth century New York changed not slowly and continuously, the way Osborn said evolution directed an organism to do, but discontinuously and through massive upheaval. The city represented all that Osborn loved and hated about evolution.

By the time Osborn was born in Fairfield, Connecticut, Manhattan was already a hodge-podge of ethnic diversity, with mansions alongside slums. In addition to the large numbers of southern Blacks and poor Whites that moved North, successive waves of Irish Catholics, Russian Jews, Poles, Italians and others came to the United States looking to improve their lives. Urban populations exploded between 1860 and 1910. Nowhere was this more evident than in New York, where many members of 'old' society saw tradition and authority breaking down. As a result, the native-born Anglo-Saxon population felt their hegemony imperiled by these newcomers. As early as the 1850s Charles Loring Brace was sounding a warning of *The Dangerous Classes of New York*. As historian Richard Hofstader characterized it, 'the Yankee found himself outnumbered and overwhelmed.'[4] It seemed to be a problem for everyone. In his influential book, *The Old World in the New* (1914), Edward Ross argued that the group that suffered most from the immigrant influx was not the wealthy or middle class, but workers. The new groups undermined wages, took scarce jobs away, and competed for the bare necessities of survival, all of which created tension. The middle class feared the working class, who in turn resented the middle class to which it aspired.[5]

Opened in 1892, by 1900 Ellis Island hit its stride as a key entry point for new arrivals. Between 1880 and 1919, seventeen million immigrants came through New York. As many spent every penny they had just getting that far, they tended to crowd into the city unable to go further. Those who stayed doubled the city's population. By 1907, for example, almost 300,000 Eastern European Jews were living in the city. This increase led to an upsurge in anti-Semitism. Whereas Catholics had been the bogeymen of nineteenth-century American politics, now the Jews took center stage in nativist nightmares.

Stereotypes began to proliferate as the specter of the stingy, conniving, mysterious and then Bolshevist agitator Jew found its way into the American psyche.[6] By 1905, 70 percent of New York's public school children were immigrants; by 1910, 40 percent of the city's entire population had been born outside the country.

Though they initially congregated in the human beehives of lower Manhattan, technology gave the newcomers a way out. The introduction of the elevator in 1904 allowed for office buildings to grow vertically, while the new subway system allowed the city to grow horizontally.[7] The first apartment building went up in 1869 and the 1870s saw the construction of the first elevated mass transit railways – the El. The city was becoming the center of business and drawing companies as well as people. Workers were commuting not just from one part of the city to another, but across the river from Brooklyn to the East and New Jersey to the West.

Along with the alien people came alien ideas like suffrage, Judaism, Catholicism, African American culture, and that most-feared threat to order, Socialism. There were two ways to deal with the problem. The newcomers could be closely restricted as to how many could enter, and as to how they were allowed to live and breed. The other way was to assimilate, or Americanize, them. This latter initiative came from both inside and outside the immigrant communities. Immigrant aid societies were often arms of various religious organizations. Churches and synagogues routinely paid for paupers and the working poor to come to America, or helped them financially after they arrived. While these organizations could sometimes transplant old world religious strife, they also helped the newcomers make the process of Americanization easier.[8] Americanization meant, as it did for such old-line patriotic groups as the Daughters of the American Revolution, not just learning English and American government and culture, but giving up connections to their old lives. Though it appeared as a liberal response to the growing immigrant population, it soon became an attempt at forced conformity.[9]

Discussing immigration, and a scandal raging over Ellis Island, Osborn explained his fundamental view on immigration to a Jewish lawyer friend, Jacob Schiff. He was not for ending immigration, but for being more selective about it. Weeding out the dregs, he said, would actually benefit other immigrants. 'I believe nothing would so greatly help our laboring class as to cleanse the competition of labor for a while.' Schiff supported the claim that William Williams, the Commissioner of Immigration, was arbitrarily excluding Jews, Germans and Italians alike. Schiff tried to get Osborn to reconsider the support he was showing the embattled commissioner. Osborn contended that Williams was an honorable civil servant being discredited by the media. At the heart of this opposition, Osborn felt, were the steamship companies (who were bringing in immigrants in the same way William Henry Osborn had enticed immigrants to the Illinois Central Railroad), and a campaign being waged 'largely from Hebrew sources.'[10] Immigration became the trigger which released Osborn's darkest feelings about his fellow humans.

Eugenics

Various groups responded differently to the wave of change coming over the country. While the Progressives sought to attack social ills through ameliorating the plight of individuals, others sought to fix society by addressing the conditions of larger social groups. The flip side of the Progressive era was the nativist and the eugenics movements. Both intertwined to a large degree in their desire to eliminate social ills by eliminating the people who suffered from them. The nativists – those born in America of Nordic, Anglo-Saxon ancestry – were deathly afraid of the large numbers of immigrants already in the country and those newly arriving. The eugenicists felt the same way, but brought to the discussion what they saw as a practical remedy. Beginning in the 1890s the nativists and eugenicists, like many Progressives, attempted 'to fulfill [their] destiny through beaurocratic means.'[11] They would apply 'scientific' techniques of sterilization and cleansing. The eugenicists also added to the basic nativist fear of foreign culture a fear of foreign biology in the form of what they called 'feeblemindedness,' as well as the fear of mixing with inferior germ plasm. Feeblemindedness was a catch-all expression meant to describe mental disabilities. Its meaning was soon made so elastic that it could be used to cover almost any social ill the eugenicists wanted it to, from simple illiteracy to destitution to severe mental impairment. In addition, miscegenation, or race mixing, was raised to the level of biological warfare.[12]

Impetus was given to the eugenics movement by the rediscovery of Mendel's laws of heredity in 1900, which seemed to support their contentions. Mendel's work suggested that physical traits were the result of 'unit characters' – single hereditary units that were responsible for single characteristics like hair color, height and such. Many researchers interested in heredity took this to mean that if a particular trait was unwanted, simply flushing that trait from the population would eliminate it. Confusing social with biological traits, this became a cornerstone of eugenic thought: to clean up a population biologically, identify unwanted traits – alcoholism, criminality, drug addiction, for example – then eliminate those individuals who carried them, or keep them from procreating or mixing their blood with those who did not have the traits.[13] Along with Mendel's work, the germ-plasm theory of August Weismann was also adopted by eugenicists. They saw it as evidence that biology alone (nature) and not environment and upbringing (nurture) was responsible for transmitting traits through the bloodstream.

The mingling of bloodlines has been a concern amongst aristocrats of the world since the beginning of recorded history. It reached the level of 'science' only in the nineteenth century. Arthur de Gobineau's *Essay on the Inequality of the Human Races* (1853) was the first book to link race and bloodlines to the study of humans. Charles Darwin's cousin Francis Galton did studies of British genealogies and determined that great men bred great men and lesser men bred lesser men. He feared the lesser orders were out-breeding the better classes. Thus, he reasoned, those of the better classes should be encouraged to breed while the lower orders should be encouraged not to. By the 1860s Galton had dubbed the concept of controlled human breeding, eugenics.[14]

Henry Fairfield Osborn's upbringing was not one which instilled in him strong prejudices and biases. His father held the standard anti-Catholic feelings of his day, but Virginia Osborn strove to teach her children to treat all people decently. While Henry Fairfield broke off his romance with Clothilde Broderotti because she was Catholic, he didn't hate her for it or think her degenerate. The one letter to Wick Scott about her is the only one in his collected papers and reminiscences where he discusses anything remotely bigoted in his early life. Beyond the typical attitudes of his age and class, Osborn did not display any overt racism before the mid-1890s and possibly not until after 1900. Even his 1891 paper, 'The Present Problem of Heredity,' was not racial and barely touched on human heredity. The 'problem' of the title was the conflict between the old and new biology. In fact, the one time that he directly addressed a human social ill was to argue that it was not biologically inherited. Using the example of alcoholism, Osborn said that the common belief that the condition is passed on to offspring through the blood 'will not bear scientific analysis' (355). He finished the paper by arguing against what would become a pillar of eugenic thought. Life would seem pointless, he said, if the 'only possible channel of actual improvement were in the selection of the fittest chains of race plasma' (363). As he believed in predestination and free will working together in evolution, Osborn would later say that 'eugenics, which has to do with being born well, and euthenics, which has to do with being nurtured and educated well, have been inseparable from the beginning of time.'[15] This is not to say, however, that he was not exposed to more extreme positions before that. As with his more mature thinking on evolution, Osborn's first real taste of racist, anti-immigrant sentiment was through Edward Drinker Cope.

Racial Lessons

Among his voluminous writings Edward Drinker Cope addressed issues of race and gender, issues he felt had a direct impact on American society because they were spinning out of control. Cope saw increased immigration of 'lower orders,' and the increasing freedom and political power of women and Blacks as undermining social order and stability. The growing suffrage movement particularly aroused his animosity. In a series of articles in the 1880s and 1890s, Cope discussed women, voting and marriage in detail.[16] He argued that women had not the emotional or mental stability to use the vote correctly. They gave into mental strain quickly, had nervous breakdowns too easily, and were biologically inclined to be dependent on men and so their vote would be negated. Though he thought women should have equal protection under the law, and have equal access to economic and educational opportunities so that they could expand their potentials as far as they could, he opposed them voting because casting votes would hinder their development. He argued that the votes of women and Blacks would be exploited by corrupt and unscrupulous politicians. In his view, 'what America needs is not an extension, but a restriction of the suffrage.'[17]

In 'Two Perils of the Indo-European' (1890) he laid out his fears.[18] Believing reformers were doing more harm than good, he wanted the government to enact a national policy on race relations and immigration so that racial strife would be averted. Attempts to bring equality, freedom and political power to women, Blacks and immigrants as a way of ameliorating their plight was sentimentalism and misguided guilt gone horribly awry. The 'peace-societies and socialists' who advocated such things were ruining the country. Like the wealthy and privileged before and after him, Cope saw full democracy as dangerous to liberty, and believed that only an élite few really possessed the ability to appreciate and use civil liberties. Anything else upset God's order. Emancipation of Blacks and their suffrage was an example to him of 'the folly which results from the endeavors to conform an abstract theory of right without regard to natural conditions' (2052).

Race mixing also went against nature's designs. Whites, Cope said, had finer faculties which would be wrecked by infusions of degenerate black blood. What was worse, miscegenation affected Whites' very souls, rendering 'the possibility of resurrection ... doubtful if not impossible.' Blacks were completely unable – spiritually and mentally – to run a government. They had not the intellectual capacity for self-rule, and all the education in the world would not change that. A black man, Cope argued, could not change his 'mental skin' any more than he could change his physical skin. Despite this, Cope said, Blacks and other inferiors should be given as much education as they could handle. What would be best, however, would be to repatriate the Black back to Africa 'where he rightly belongs' (2054). In addition, not only did Cope want to reform 'normal marriage,' but also to outlaw such unions between alcoholics, drug addicts and the insane.[19]

Besides miscegenation and suffrage, another of the 'perils of the Indo-European' Cope feared was the breakdown of the boundary between masculinity and femininity. He said the increasingly easy life of many white men in America was making them soft. Too much of this rarified atmosphere over a few generations 'may very probably emasculate men, so that their true sex-qualities are almost entirely overgrown.' Men who supported women's suffrage, he claimed, were 'effeminate and long haired.' He concluded by saying, 'let us have no more feminine men and masculine women' (2070–2071). Edward Drinker Cope was not the only one who would help draw Osborn to the dark side of anthropology.

While Osborn was associated with many leading members of the eugenics movement – Charles Davenport, Harry Laughlin, Lothrop Stoddard and others – his closest and most zealous eugenicist friend was the New York lawyer and amateur naturalist, Madison Grant, the author of two of the most outrageously racist works in American letters. The published work of Madison Grant (1865–1937) has been considered by historians to be an important part of the eugenics movement because of its role in spreading eugenic notions throughout the United States and Europe. Virtually every book that deals with the history of eugenics mentions or discusses Grant's work to some extent.[20] Grant believed three things: that white Anglo-Saxon Protestant, Nordic males (himself included) were the ultimate expression of human evolution; that the United States was under biological attack by racial inferiors; and that something had to be done to stop the onslaught. He used eugenic arguments to support his openly supremacist ideology in an attempt to rouse the Nordic population of the New World to the danger, and rally them to take action.

Madison's father, Gabriel Grant, was born in Newark, New Jersey in 1826. He studied medicine in both Newark and New York City, where he graduated from the College of Physicians and Surgeons in 1851. After organizing the American Hospital in Panama in 1852, he returned to Newark to practice. During the cholera epidemic of 1854 he was appointed a member of the Newark Health Commission. At the start of the Civil War he joined the Second Regiment of New Jersey Volunteers under Phillip Kearny, as a surgeon. He saw action at Bull Run, Gaine's Mills, Antietam and Fair Oaks, at which he won the Congressional Medal of Honor. He was eventually appointed director of the government hospitals at Evanston and then Madison, Indiana. He ran the latter institution until he left the Army in 1865.[21] Gabriel Grant was also a mason who once gave a speech lauding George Washington as a 'champion of human rights … a rallying cry of nations and the oppressed masses of Europe.'[22] The reason that Masonic principles were important to the country, he said, was because within the craft 'arbitrary political distinctions were nothing: All were equals and brothers.' Madison dedicated his first book to his father.

By 1881 Gabriel Grant had moved his family to New York City. Little is known of his son's early life. He seems to have consciously expunged a good deal of it from the historic record, and kept the rest from view. Once

Fig. 6 Madison Grant. Zoo builder, highway maker, political wheeler-dealer, and Osborn's friend. The author of arguably the two most racist works in American letters.

Neg. No. 2A5186, courtesy Department of Library Services, American Museum of Natural History.

asked by publisher Charles Scribner to supply personal data for a promotional flyer, he sent the bare essentials, only saying, 'I don't like to talk about myself and have kept it in small amounts.'[23] Born November 19, 1865, Madison Grant's early education consisted of a private school in Dresden, Germany. He then attended Yale (class of 1887), and finally Columbia, where he received a law degree in 1890. The family wealth allowed the boy to travel extensively throughout Europe, Asia and Africa. On these trips his awareness of race grew as he observed exotic cultures directly.[24] Settling into the life of a New York socialite, he became involved in immigration and conservation issues.

While he did not talk about his immediate family history, he was exceedingly proud of his supposed Nordic ancestry. Grant disliked non-Nordics (and especially immigrants) to an astonishing degree. To him they were vermin intent on destroying him and his world. He thought they had not and could not contribute anything to American society. He took the position that anyone who had contributed anything to the country's advancement must have been an Anglo-Saxon because no other group was capable of such things. This belief led him to assert that Jesus Christ had been a Nordic.[25] He generally used the terms Nordic and Anglo-Saxon interchangeably, and Aryan only when discussing human origins. He once scoffed at a planned memorial in Battery Park, New York celebrating the contributions of immigrants, and tried to stop its construction. He told Osborn, 'we should do everything possible to kill this glorification of the alien.'[26]

Madison Grant's avocation was natural history, though he had no formal training in science.[27] Despite his lack of appropriate credentials to do so, he often referred to himself as a scientist. He was an avid hunter and explorer. His involvement in the study of North American mammals resulted in one being named for him: the Alaskan caribou, *Rangifer granti*.[28] In 1894 Grant's enthusiasm for zoology led him to take a leading role in the creation of the New York Zoological Society Park (the Bronx Zoo) with Osborn, Theodore Roosevelt, future Nobel Peace Prize winner Elihu Root, and the grand old man of New York politics, Andrew H. Green (a longtime friend of Osborn's father William Henry). When they met – probably in 1894 – Osborn and Grant (eight years his junior) took an immediate liking to one another. Osborn thought enough of Grant to want him in a place of power at the American Museum as well as the Zoological Society. He had to do it over the objections of Green, the man who had helped write the museum's original charter. It was not easy. 'I had a big fight with Andrew H. Green,' he told Fairfield Jr, 'to get Madison Grant on the [museum] executive committee.'[29] Grant and Osborn became close friends, dining and meeting on a regular basis. They held a great respect for each other's goals and abilities. They were so close that William Hornaday said they had become 'welded into that one purpose [of promoting the park and conservation issues] like the best team of horses.'[30]

Andrew H. Green (1820–1903) was a powerful New York public official and preservationist. He was president of the Central Park Commission from 1857 to 1871 and city comptroller during the break-up of the corrupt 'Boss' Tweed, Tammany Hall political machine which had such a stranglehold on New York politics. In 1868 he began the process of bringing the five boroughs of New York together as a unified political entity controlled out of Manhattan. The process of consolidation took effect in 1895, creating 'Greater New York' as a megalopolis of five million inhabitants. He was also instrumental in creating the Niagara Falls nature preserve and the American Scenic and Historic Preservation Society.[31] The man called the father of Greater New York disliked Grant as much as Grant disliked foreigners, and resisted his installation on both the park's and museum's boards.[32] In response, Osborn and Grant engineered Green's ouster from the park's development apparatus.[33]

For the men who built it, the Zoological Society and Park had great cultural and political meaning. Through various programs and public relations, the zoological park was intended to become a vehicle by which Grant and his associates could promote the ideals of their elite group. Showing the flair and agility of a politician, Grant involved himself in the New York mayoral campaign of William Lafayette Strong, behind the leadership of Republican boss Tom Platt. The reformers wanted to rebuild the largely Irish police and fire departments as well as to undermine Democratic power, which was built on poor and immigrant constituents. Upon Strong's 1894 victory, he gave Grant the authorization to build the park. Since the zoo was in a then inaccessible part of the Bronx, Grant suggested a highway be built, and used his connection with Theodore Roosevelt (by then commissioner of police) to get himself on the planning board. The resulting Bronx River Parkway became the first major parkway in the United States.[34]

The plan for the park – the founders hated it being referred to as the Bronx Zoo – was hatched at the exclusive Boone and Crockett Club of Manhattan. The club had been formed in 1887 by Theodore Roosevelt and George Bird Grinell as a watering hole and meeting place for the city's élite big-game hunters. (You had to have killed a big-game animal with a rifle in order to be made a full member.) The rich, big-city huntsmen were outdoor and natural history dilettantes who wanted, among other things, to preserve their vision of nature for people like themselves, and to protect it from what they saw as unappreciative mobs of immigrants and rapacious corporations who, the members thought, were destroying it. Roosevelt had become concerned about preserving wildlife, and the growing decadence of the country. He would later voice concerns over 'race suicide' as well.[35] Living his 'strenuous life,' Roosevelt embodied the popular form of Osborn's struggle hypothesis. Grinell was, like Madison Grant, a New Yorker, lawyer, and Yale graduate. He had gone on the early fossil hunting trips out West with O.C. Marsh in the 1870s.[36] Osborn had been introduced to the New York preservationist circle by Roosevelt when he returned to the city in 1891 to accept the positions at Columbia and the American Museum. The two had known each other since their childhood frolics on the Hudson River, and were now getting re-acquainted. Though not really a hunter, Osborn was made an honorary Boone and Crockett Club member in 1899.

Grant took credit for putting forward the initial idea of the park to the other members of the club.[37] He was inspired by the zoo being created by the Smithsonian Institution in Washington. The park board then invited the man responsible for the government zoo, William Hornaday (1854–1937), to come away from the Smithsonian to help design and direct the New York facility. Hornaday fit in well with the other Zoological Society members. He was passionate about conservation and had little good to say about immigrants, feeling they were a threat not only to wildlife, but also to the country.[38]

The founders wanted the park to reflect their views on how nature, and thus human society, should exist. It was to be a place where they could enjoy the manly outdoors and the spiritual awakening it brought them. It was also to be a place where the masses could come and learn (under strictly controlled conditions) the lessons of nature and how those lessons should be applied to everyday life. The park was laid out to teach immigrants and the native poor how to behave in public places and how important the preservation of nature was. Littering, cutting trees for firewood and bird shooting were strictly prohibited and enforced through signs, guards and newspaper articles.[39] This idea of the wealthy few teaching the starving masses was apparent to some. The *New York Sun* challenged it, saying the park was going to be a 'toy' for the rich founders to play with.[40]

Once the zoo was completed, the people came. The growing availability of automobiles made the city's population more mobile. Madison Grant's Bronx River Parkway, originally built to allow his friends easy access to their nature-boy paradise, also allowed the very people he hated most easy access to the wilderness he said they did not deserve. Not long after, Yankee Stadium was built, drawing more people to the area. The new arrivals began moving out of the Lower East Side, and uptown into more northerly neighborhoods – not just for brief forays, but to live permanently – getting disturbingly close to the environs of the middle-class Yankees like Osborn who complained of this mobility. He said that 'thousands of East Side Jews' were flocking to and clogging up his favorite bucolic haunts.[41]

Conservation

As chairman of the New York Zoological Society, Madison Grant was deeply involved in the park's operation as well as conservation issues in general.[42] He was so involved that the *New York Herald* said his various projects 'were better called conservationist then conservative,'[43] and was once referred to as the 'godfather' of conservation efforts in America.[44] Grant also helped found the 'Save the Redwoods League' in 1917 to preserve old-growth forests from lumbermen. Though usually only acknowledged as such by historians, Grant was one of the founders of the modern conservation movement. Preserving important wildlife, especially the larger mammals and birds, was high on Grant's list of priorities. His efforts in conservation issues reveal a man deeply concerned with the state of wildlife, not only in the United States, but around the world. It also gives insight into his racial views. In the inter-war years in Europe, for example, large game was being hunted to the point of extinction. The problem was that peasants now had guns, with which 'they shoot anything and everything.'[45] Grant was outraged at the notion that rare, majestic species were disappearing at the hands of the unwashed, degenerate rabble.

There were pressing questions of conservation closer to home as well. In the early 1930s, the height of the Great Depression, wildlife was under attack in the State of Washington. Grant had learned that starving homeless people had taken to the forests in search of game to kill and eat. He feverishly contacted his friend Senator Albert Johnson, the Washington State Republican, to see if it was true. Johnson verified that it was indeed so. Grant asked if some legal action could be brought to bear upon the problem. Johnson was politician enough to realize the bad form of keeping people from eating the game. While he agreed with Grant, he cautioned him against looking as if he favored wildlife over starving Americans. Keeping out immigrants was one thing, but this was something else.[46]

After his success with the park, Grant advanced his status in the world of racial activism in 1909 when he became vice-president of the Immigration Restriction League (Osborn was also later a member). The league was started in 1895 by Prescott Hall and Robert DeCourcy Ward, a pair of Boston brahmins who as Harvard students became interested in immigration. With Henry Cabot Lodge as a spokesman, they lobbied for mandatory intelligence tests for incoming immigrants. Like Grant, Hall was a lawyer who practiced little, but instead threw himself into immigration and eugenics issues, writing *Immigration and Its Effects on the United States* in 1906.[47] Grant was not the only one with a growing sense of the activist spirit.

In his evolution work Henry Fairfield Osborn connected the physical and the spiritual. Advancement in one without the other was no advance at all; both had to grow or the species was doomed. Since he believed that history as written in the fossil record had proved this equation a reality, he grew concerned that his fellow human beings were not living up to the challenge to advance as far as they could. They were suffering physically and morally. It was bad enough that the faceless sea of humanity was faltering, but so were people around him. It saddened him when he found out that an acquaintance,

Poultney Bigelow, had been caught in an extra-marital affair. When the news broke he told Loulu: 'it all confirms my ever increasing conviction that without religious belief and morals the world will rapidly go to Hell.'[48]

If he was concerned over the moral well-being of the adults around him, he was even more so over that of his children. Once again Josephine had a special place in his heart. She was his 'chip off the old block' and he was concerned about her future in a world that had changed so radically from the one he grew up in. While she was away at camp in Maine or on other travels he missed her a great deal and worried over the fact that she was growing up so fast and would soon no longer be his little girl. The list of things that concerned him was growing longer, and he was growing more public in his opposition to them.

In addition to moral cleanliness, physical cleanliness was of paramount importance. Opening an exhibition at the museum on public health gave Osborn a chance to voice his concern over the medical condition of the city of New York. He laid out his tough-love approach to the realities of unrestricted immigration and its attendant health problems. He took to heart the adage that cleanliness is next to Godliness, and compared how nature cleaned up after itself with how people did. Nature was the clear winner because nature followed God's rules. If humans did the same there would be benefits for all. 'It is certain,' he said, 'that in the advance of civilization the more close we get to nature's laws and example, the more civilized we shall be.'[49] To Osborn, the fact that so many children lived in abject squalor and poverty in the greatest city on earth was tantamount to child abuse on a societal scale. That even more children were being born into such conditions while more entered every day through immigration was a national shame. These unfortunate children, he said, were 'predestined to disease and suffering.' The creation of these swarms was unfair and cruel not only to the new arrivals, but to the children already here. 'It is cruel to bring children up in an unclean environment,' hence there must be a concerted effort to clean up the slums of New York, educate and clean up the children living there, reduce the numbers of their births, and carefully select the new immigrants coming in. The museum represented one of the best ways to teach the city's children the importance of nature's lessons, and so he lobbied hard for city and state funds to expand the museum as a center of learning for the city's millions.[50] Like the ghost of Christmas Present in Dickens's novel, Osborn saw the greatest threats of the Industrial Revolution as the children, Ignorance and Want, who clutched at everything and ruined it with their dirty little fingers. As the new century progressed he became more outspoken and active in his efforts to combat what he saw as the growing evil of social and moral decay. That evil was about to grow larger.

The Great War

The start of World War I in 1914 seemed to usher in a dark and foreboding future that made a joke of ideas about a stable, optimistic, progressive world. The Great War held significance for Osborn and his fellow eugenicists beyond that for other people. For them it was not just a terrible war that could destroy European civilization, spread around the world, and engulf the United States. For Osborn, Grant and the others it was bad because, as they saw it, mostly Nordics would die, thereby reducing their already depleted numbers. Eugenicists believed that such human 'traits' as courage, honor and duty were strong in Nordics and weak in other groups. The highly motivated Nordics, they argued, would do most of the fighting and dying while other racial groups sat out the conflict and survived. This made the war 'dysgenic:' bad for the Nordic race on a biological level.[51]

The war for Osborn had a concrete as well as an abstract consequence. On the eve of the conflict Henry Fairfield Jr was in Europe arranging his marriage to an English woman, Marjorie Lamond. They wanted to get married on the continent and were searching for a suitable place to hold the ceremony. When the fighting broke out the families lost track of them for several tense days until Fairfield Jr and his bride-to-be turned up in England.[52]

Osborn was staunchly pro-British and felt if war did come the Germans would be responsible. In August of 1914 he wrote to Edward Poulton to tell him that those who mattered in America were squarely in the British camp. He also suspected that the masses and those in control of the media would not show the right attitude towards the situation. When the war began Osborn was surprised to admit that non-Nordics in America were getting into the proper mood. With a backhanded compliment reminiscent of the one his father had made about his scientific career, he said that 'even the Jews of German origin, who now control two of our greatest newspapers, talk like Anglo-Saxons.'[53]

Unfortunately, Poulton's son, Robert, was killed the next year in France. Osborn told Theodore Roosevelt about Poulton's loss shortly after hearing the news. Roosevelt, already worked up about the European situation, wanted America in the war in the worst way. Roosevelt had once been a supporter of German militarism, but now opposed it vigorously. He blustered to Osborn, 'You tell Poulton for me that that if I were [still] president we would be shoulder to shoulder with the allies.'[54] H.L. Mencken barbed Roosevelt for his vacillation by saying that 'such violent swallowings and regurgitation's were no novelty to the man.'[55] Unlike Roosevelt, Osborn was not so sure about the United States getting involved. It would undermine America's moral high ground if her troops were sent to fight. He believed, like Woodrow Wilson, that America should work to end the war diplomatically.[56] By the end of 1917, however, his attitude had changed. With America mobilizing, he got on board. He was proud that so many of the museum's staff volunteered, though he was worried that the draft would take only Nordics, and not 'all classes.'[57]

The Osborn family played their part in the war effort. Osborn himself was a committee member for the New York office of the United States Food Administration. Perry served as a Lieutenant-Colonel on General Pershing's staff. Fairfield Jr was a captain of a 'colored' artillery company, and Virginia and Josephine went to Paris as aid workers. Virginia in particular had much to burden her. She had married Ralph Sanger in 1904, and her first-born son Osborn died tragically young; Ralph was serving as a flyer.[58] There were, however, ways some sought to serve their country other than wearing a uniform.

The Great Race

Seeing the world crumbling around him, and believing he had the answer to the crisis, Madison Grant developed what he thought was a grand idea for a book. Grant and Osborn had formed the Half-Moon Club in 1906 for New York aristocrats with a taste for adventure. They met several times a year for overblown black-tie dinners. The attendees were referred to as Mariners and Pilots. The highlight of each dinner was a special guest lecturer. In 1908 the guest was William Zebina Ripley, the Columbia University economist turned anthropologist. His 1899 *Races of Europe* was a monumental 624-page tome on the racial history of European populations and the dangers of the lower orders, especially the Jews. The book's major contribution to anthropological research was to separate Caucasians into three distinct sub-races; the Teutons (or Nordics); the Mediterraneans; and the Alpines. Though Grant was already interested in anthropology, argues author Jonathan Spiro, Ripley's talk galvanized him to action. Where Grant had once studied the downfall of noble animal species, he would now do the same with humans.[59]

He quickly put together a rough draft and went looking for a publisher. Grant knew that Osborn and publishing giant Charles Scribner had been friends since their Princeton days, and talked to Osborn about the manuscript. Riding the crest of his success with *The Men of the Old Stone Age*, Osborn approached Scribner about Grant's book and suggested it be published. Not only did Osborn write the glowing introduction to the work, he revised the final draft.[60]

With the war raging Grant was eager to get his book in print. When it did finally appear in 1916, *The Passing of the Great Race* generated little notice. By the 1920s, however, it had become the central work of the immigration-restriction cause. The book is a retelling of Western history along racial lines with Nordic supremacy as its central thread. According to Grant, Nordics are the great race being pushed out and overrun by 'lesser breeds' and human mongrels. After discussing the Central Asian origins of the Nordic race, Grant went on to explain the dangers and deprivations being heaped upon the Nordic world – particularly the United States – by Jews, Blacks, Eastern Europeans and other 'undesirables.' He recommended such remedies as the controlled breeding of humans in order to remove degenerate traits, abandoning democratic ideals such as the 'brotherhood of man,' instituting the forced sterilization of certain populations, and bringing to a halt the uncontrolled immigration of the unwanted into the country. All these things, Grant felt, were 'sweeping the Nation towards a racial abyss.'[61] Grant subscribed to the basic tripartite plan held by most eugenicists for dealing with this growing biological threat: first, keep out as many immigrants as possible; second, segregate undesirables already here to keep them from mixing with the native population; and third, sterilize them so they could not breed more of their kind.

Madison Grant's frantic racism had thus found scientific focus. Osborn's Central Asia hypothesis gave Grant a starting point for his history of the Nordics, the foundation of *The Passing of the Great Race*. Working closely with Osborn on the zoological park, Grant imbibed Osborn's theory of human evolution. He also found theoretical support for his beliefs in the works of Arthur de Gobineau as well as in those of William Zebina Ripley. With these works as 'authorities,' Grant was able to dress his racism and anti-Semitism in a cloak of anthropological respectability.

In his short introduction, Osborn was full of praise for Grant, whom he saw as building a biologically based history of Europe. Osborn said Grant's work was important because he explained how race had played a greater role in European history than either language or nationality 'in molding the destinies of Men.' Though he held Nordics up as the possessors of the best characteristics in people, however, Osborn was still reluctant to dismiss the rest of the human race the way Grant was. He said that America must cultivate the best in all groups. 'Divine forces' appear in all people, even 'in what we call the lowest races.' Conservation of these positive traits was paramount for America's survival.[62]

Whereas Grant drew from Osborn the latter's science, Osborn drew, consciously or not, from Grant's hard, angry racism. Osborn had already adopted a basic Nordicist outlook, but he saw the issue in terms of neutral science – at least that is what he told himself. Whatever his personal feelings about race relations were, he did not openly discuss them, in public or private, before he came back to New York and become part of the conservationist, nativist movement there. Once he did become vocal, Osborn was caught in a trap of his own making, trying to play the role of scientific objectivist – to keep the warring sides friendly, to line up facts so that they spoke for themselves with the imprimatur of God – he could sometimes have trouble controlling his less generous feelings.[63] The dinosaurs and cavemen had followed nature's laws and prospered, and so could modern humans. It seemed so easy and obvious. He needed to convince others he was right. To that end he became a eugenics activist, and involved himself in conservation and immigration issues.

The Roaring Twenties

New York City had changed so thoroughly that in 1910 it bore almost no resemblance to the city that had existed in 1880. Modern New York had appeared and Osborn's beloved old New York was gone forever, lost in a surging sea of new faces, foods, religions and languages.[64] By 1920 he was lamenting that New York had become a 'decadent social environment,' making it difficult for young people to find their way.[65] That year one of the central symbols of his ordered venerable world disappeared: he sold 'dear old 850 Madison Avenue.' It had been his home for almost thirty years. The house had echoed with the sounds of his children and his adoring wife, it had been the site of uncounted holiday celebrations, parties and dinners, and played host to many of the leading scientific, political and social leaders of the day. Even though he was now sixty-three years old and feeling his age, even though the children were all grown and the mansion too large for just two people, and as such not economic to live in, he was still heartbroken to lose it.

After selling the house, Henry and Loulu moved to an apartment on an upper floor of 998 Fifth Avenue near the Metropolitan Museum of Art. To add insult to the injury of leaving the house, Osborn had to sue the real-estate agent, the Morganthau Company, claiming they misrepresented their intentions for the property and at the last minute had tried to change the sale agreement. The building was eventually torn down to make way for an apartment complex.[66] It was an ominous way for Osborn to begin the 1920s.

His children were also getting on with their lives. Virginia stayed at Castle Rock with her father and tried to drown the sorrow of the death of her child, and of her husband Ralph, killed in the war, by tending the gardens.[67] Following her war service, Osborn told his youngest daughter Josephine that she was now 'the master of [her] fate.' He was very proud of her, yet was concerned about the world she was living in. He confided to her that even happiness came only from hard work. 'We Osborn's,' he said, 'have to work to be happy – it is our genius.'[68] Josephine, who had stayed behind in Europe in order to revive her social life, was now gleefully on her own. One of her favorite places was St Moritz. At thirty-two, she had still not married (something her father saw as very important to race survival). Feeling no hurry to wed, Josephine told her mother that 'I have the funny disposition of being happy where I am.' Her lack of a husband was in part a result of having a father of such fastidious moral bearing. As Henry Fairfield Osborn could not imagine marrying a woman who was not at least his mother's equal, Josephine Osborn could not marry a man who did not measure up to her father's standards. The men of St Moritz and other European glamour holes drank too much, womanized too much, and exhibited an endless line of 'hopeless qualities.'[69] She eventually found a man worthy of her in Jay Coogan. They married in 1926 and quickly produced two children. After all that, her father still called her 'our darling baby Josephine.'[70]

Both of Osborn's sons – Perry, now a lawyer and Fairfield Jr, an oil company executive – developed an interest in eugenics. Fairfield Jr was the more vocal of the two: he once scolded his father for being a committee member of the American Indian Defense Association, and told him to quit because it did not look good for a man like him to be a part of such an organization. Fairfield went on to write several eugenics books of his own and become an active member of the movement.[71]

For Osborn and his milieu, the crisis extended from intellectual to cultural change. The 1913 Armory Show introduced modern art to the wider public and Socialism was becoming a rage large enough to require a crackdown by the authorities. What Osborn's fellow Scribner's author, F. Scott Fitzgerald, called the 'roaring twenties,' his other fellow Scribner's author, Lothrop Stoddard, called 'the revolt against civilization.' (Fitzgerald parodied Stoddard, and probably Madison Grant as well, in *The Great Gatsby*.) With the war over and America seemingly on top of the world, a new social sensibility swept in. A new freedom of behavior left strict moral codes behind. Women were growing more independent and adventuresome. The long struggle begun by Elizabeth Caddy Stanton, Susan B. Anthony and others finally saw fruit in 1920 with the passage of the Nineteenth Amendment giving women the vote.

Carrie Chapman Catt (1859–1947) had taken over the reigns of the women's voting rights struggle and saw it through to completion. She rallied support from many different quarters, including the president of the United States. The president of the American Museum of Natural History was a much tougher sell. Catt wanted Osborn's support for the final push of the amendment, but he declined. While he supported the general idea of women getting ahead, he refused to give his public support to Catt because, he said, voting would not 'favor the evolution of women.' Like Cope, Osborn felt it would only complicate their lives and make it tougher for them to attain spiritual advancement. He wanted evolution, not revolution. As with many issues, he left himself a way out by telling Catt that though he felt the way he did, 'I am of course open to a change of mind on this point.'[72]

Women's suffrage was not the only culture shock hitting New York City. It was the era of the 'New Woman' and the 'New Negro.' Starting around 1914, large numbers of Blacks began to move uptown to Harlem and to turn the neighborhood into a center of African American culture, art and literature. By 1920, Harlem had become 'Harlem.' The most obvious form of black culture, jazz, swept white America by storm. The music, dress, language and attitude of Black culture became popular and did more to integrate the country than most relief efforts ever could. Alongside the flowering of the 'Harlem Renaissance,' White culture also grew bolder. Prohibition, enacted in 1919 to save American youth from the horrors of moral decadence and social breakdown, only hastened it along by making drinking a delicious and stylishly dangerous sport. It also allowed for the first appearance of organized crime and the advent of the gangster era. Gangsters became fashionable to the smart set, and political heroes to many of the downtrodden. Speakeasies flourished and became places where the social élite of New York could drink illicit whiskey, listen to Black jazz groups, and feel very hip. Society types stayed up late and got up bleary-eyed. They did not work at work; they worked at having fun and had no moral problem with their behavior. Osborn and his friends simply thought they had no morals.

Henry Fairfield Osborn, Madison Grant and the other eugenicists and immigration restrictionists were preoccupied with the idea that America was being mongrelized by these new conditions. A mix of low lives and dangerous germ plasm was a condition that spelled the doom of the country. For many of the up and coming intellectuals of the era like Dorothy Parker and her circle of friends at the Algonquin round table, being a mongrel was a positive thing. Like Emerson and Walt Whitman before them, they wanted to break away from European convention. To the Moderns, a mongrel was a mixed breed incorporating many different backgrounds and sensibilities to create something better, something more honest. They were suspicious of the blue bloods. Parker, though she never actually wrote it, planned to call her autobiography *Mongrel* as a sign of pride, rather than derision. She and other writers of her generation chose to satirize and poke fun at the hypocrisies of Osborn's self-important, sanctimonious world. She attacked upper-class paternalistic attitudes towards Blacks. In her 'Arrangement in Black and White,' a rich and clueless party-goer describes her husband's views of Blacks in a way reminiscent of Osborn and Roosevelt. 'All he says is, he says he hasn't got a word to say against colored people as long as they keep their place.'[73] As earlier generations of writers and intellectuals sought salvation in the bucolic splendor of the American landscape, Parker and her friends looked for it in the gutters and alleys of Manhattan, and in what Ann Douglas has called 'terrible honesty.'[74] People were experiencing a wave of 'rampant individualism – not only in art and literature,' Osborn claimed, 'but in all our social institutions.' It was an individualism that Osborn said threatened 'the very existence of the family,' and hence American society itself.[75]

Public Man

As president of the American Museum, Osborn had a ready-made platform from which to address the public on different issues. The Gobi Desert expeditions of the 1920s gave him and the museum name recognition and he capitalized on it in order to bring people's attention to social concerns he thought important: race relations, eugenics, immigration, social order and, most importantly, the study of science and evolution. He always chose issues which he saw as having a direct relationship with the effects of evolution. He had published a few scattered articles before the 1920s, but then began to increase his output of more overtly political pieces. Osborn thought he too had the answers to all the country's ills. He lectured public school teachers not only on how important teaching science and evolution was, but he showed them how to do it.[76] He then told them the proper way to carefully 'select' children who would do well.[77] He even began giving examples of people he thought embodied his theories of inborn greatness. He claimed that the world's leading naturalists were such men because of their 'racial souls.' Osborn claimed that all races had their racial soul, which was 'the spiritual, intellectual and moral reaction to environment and our daily experience.'[78] Men such as Darwin, Huxley, John Muir and Theodore Roosevelt had an inner drive and ability to tap into their creative genius that gave them an intellectual and spiritual edge.[79] These men were 'naturalists,' not scientists. A naturalist learned from nature (and thus God), while scientists learned in the sterility of the laboratory. Osborn said he had studied not only the physical, but also the spiritual growth of early man. This study gave him insight into the problems facing modern society and how to remedy them. People were losing their creative, spiritual edge. He railed against the tabloid press, against people who did not properly raise their children, did not carefully choose their mates, and against a breakdown in the appreciation for the finer, more cultivated, uplifting things in life. He predicted that the recently formed Soviet Union would crumble because it did not promote close family ties, or civic virtue, and added that the United States could suffer the same fate. The secret was this: 'man must not only think of the present ... but of the future.'[80] The problem of the future was 'not more people, but better people.'[81] But he knew his ideas were not popular and that he was being thwarted at every turn, and so his anxiety grew. In philosopher Michael Ruse's words Osborn was suffering from 'the worries of a public man.'[82]

In 1918 Osborn, along with Grant and biologist Charles Davenport, formed the Galton Society. It was begun as an organization to study human anthropology, evolution and eugenics.[83] They had become worried that mainstream anthropology, especially as it was now being so permeated by the work of Franz Boas, the pioneering Jewish American and Columbia University anthropologist, was not giving eugenics its proper due. Boas introduced the relativist approach to the study of culture, and championed the role of environment and upbringing (nurture) as more important in the development of human society and potential than biology alone (nature). Boas's work was a direct assault on everything Grant and his eugenicist cronies held dear. It was also a bit too radical for Osborn's tastes. He did support the notion that a person could uplift themselves by personal effort and certain environmental conditions, but also held for a powerful predestination as well. If biology counted for little, as Boas claimed, then Osborn's Nordic heritage and race plasm meant nothing. If they didn't, then his entire theory of evolution, the thing he built his career around and based his chances for salvation on, also meant nothing.

Although he was a Nordicist, Osborn claimed that he did not make it a special study or preoccupation despite his growing pontification on the subject. When asked about the racial landscape in America by the British Ambassador, A.J. Balfour, Osborn said that the Northeast 'is thoroughly American in every sense, but is perhaps less optimistic in reform measures and more cautious in new legislation, suffrage, prohibition, and questions of that kind.' Seeming to know quite a bit about the subject, he said that many men in the East 'retain more or less of their racial characteristics,' and represent 'Old world racial types modified by American ideals.'[84] When asked to give his opinion of the eugenics book *Human Behavior*, by Stewart Paton, Osborn said he liked it, 'but the general subject is so far out of my field of work that I am unable to form any opinion of its value.'[85]

In 1914, in an effort to combat the rising problem of social ills, American eugenicists put together a major gathering of like-minded researchers from around the world. The Second International Congress on Eugenics was the brain-child of Charles Davenport, Ales Hrdlicka, Vernon Kellog, and others. They had attended the First Eugenics Congress in London in 1912 and wanted to bridge the gap between American and world eugenics organizations. They began organizing their own congress for 1915. Charles Davenport wanted Osborn to preside, but he was reluctant to do so. He told the organizing committee that he would only do it if the congress was held at the American Museum, so that he would not have to be overly involved. Osborn would only attend the opening meeting and would not help run the show.[86] The war interrupted the plans, but the committee vowed to resume when the hostilities were over.[87]

Filled with a new urgency following the war, the organizers constituted the Second International Eugenics Congress. Again, Osborn was asked to be president and the meeting was held at the museum. Davenport wanted the gathering to be truly international as a way of bringing the American and international eugenics communities back together. He was particularly interested in bringing 'certain individual eugenicists in Germany' back into the fold.[88] Internal dissent grew over the German question. Some countries' delegates thought the Germans should be kept out because of Germany's part in the war. Herbert Spencer Jennings wanted the Germans included. He said it would not be truly international without them. Jennings complained that delegates were invited not based on scientific criteria, 'but on grounds irrelevant to science.'[89] In the end, Osborn bowed to the anti-German sentiment.[90] Despite Jennings's qualms, the meeting did bring together an international list of eugenicists. Speakers included William McDougall, Edward East and Ernest Hooton of Harvard, Raymond Pearl of Johns Hopkins, Stewart Paton of Princeton, delegates from France, Italy, Norway, England, India, Scotland and Cuba, as well as representatives from the Metropolitan Life and Prudential Insurance Companies.

Osborn said the congress brought together researchers to study whether the United States would build up its own civilization or let some other country or government do it for them. The congress was to be a call to government action on immigration and eugenics education. He believed in 'the right of the state to safeguard the character and integrity of the race.' The congress would show that education alone, while being able to improve individuals and even entire races, could not fundamentally alter the deeper race plasm.[91]

To all involved the congress was a major success. They even received kudos from some unexpected quarters. Archbishop Hayes, the Catholic prelate, praised the congress and agreed with many of its aims.[92] Osborn, buoyed by this support, agreed with Yale economist Irving Fisher that a Jewish and Catholic representative should be involved with the next congress – as long as they were 'broad minded.'[93]

Elated by the congress's success, Osborn went to the newspapers with renewed zest. One of his more outspoken eugenics articles appeared in April 1924. Titled 'Lo, the Poor Nordic,' it was meant to address the current condition of 'my own race, the Nordics.'[94] He discussed how Nordics had first appeared in Central Asia and then made their way into Europe about 12,000 BC. Their struggle to overcome the adversity of their environment, helped along by their race plasm, made them superior. They eventually gave rise to all the Europeans of distinction from Da Vinci to Cervantes, Giotto, Galileo, Petrarch and Columbus, before crossing into North America to find their ultimate destiny. He finished by throwing a few crumbs to the lesser breeds by saying that all races produce great men, but each race should celebrate its own, not 'men who really belong to other races.' Though he may have missed the irony of gobbling up all the great men for himself, leaving none to others, readers did not. A few days later the editorial pages erupted with several angry letters, led by one from Franz Boas, rebutting his position. In measured tones Boas dissected Osborn's claims and concluded that it was 'an old error to judge history from the present.'[95] Another writer pointed out that Osborn had taken most of his argument from Stewart

Houston Chamberlin and his Germanic superiority writings.[96] *New York World* editor Walter Lippmann chided Osborn for his involvement in the congress and for 'lending your name to those easygoing generalizations which I am sure have no scientific foundation.'[97] Osborn was sure he could convince him otherwise. He also assured Lippmann that he knew the author was kidding. 'There is one thought I have above all others,' he told Lippmann, 'and that is to keep good humored about [matters of race].'[98] In addition, Osborn was uncomfortable with being thought of as a eugenics spokesman.

In 1924, H.B. Hayden wrote to Osborn asking advice over a sound eugenic marriage between a Nordic friend and a woman who was half-Jewish. Osborn declined to give such advice because, he said, he was not qualified to. Instead he sent Hayden to Charles Davenport, whom he considered an expert on such matters.[99] Several years later, he was still claiming to know little about the subject. In 1927 eugenicist Leon Whitney asked Osborn to read his manuscript for a children's eugenics book, and then recommend it to his publisher. Again, Osborn sent the request to Davenport, claiming not to be up to such a review.[100]

While Madison Grant's answer to social problems was to eliminate 'degenerate blood' from the American system, by force if necessary, Osborn had a more sympathetic view. He did feel that the lowest dregs of society were a problem, and that America had been made a dumping ground for Europe, but he also opposed blanket restrictions based upon race only.[101] He wanted 'not more emigrants, but better emigrants.'[102] Though Osborn believed that the races were not all equal anatomically or spiritually, he did believe that there were quality non-Nordics who could contribute to American society and make good citizens and that the improvement of all races benefited the country.[103] He argued the fine line between selection and restriction, favoring the former. All the races were capable of improving themselves – if they could not, how could he? There were good Jews and Catholics, good Poles, Italians, Slavs and Asians. They should be allowed in while alcoholics, criminals, Bolshevists and other degenerates should not. He wanted to cultivate the best from all races for the American utopia he envisioned. He argued that 'there are very good and desirable emigrants to be found in every country.' Thus he admonished Madison Grant for having too narrow and biased a view, a view Osborn felt was detrimental in the long run.[104] He was also troubled by an apparent rise in the number of successful immigrants who, having done well in the United States, went back to their home country to ply their newly acquired skills there, rather than here. The exit of desirable immigrants was just as big a problem for Osborn as the entry of undesirable ones.[105]

He did agree completely, however, with Madison Grant on the issue of miscegenation, or race mixing. The melting-pot concept was not good, he argued. The races should not be mixing. Those who said it had no effect –

meaning Boas – were wrong because miscegenation was ruinous to any species. Osborn used the example of ancient life. The Neanderthals and Cro-Magnons did not mix, and neither should modern humans. He claimed 'the cavemen knew.'[106] The history of living things, especially humans, Osborn argued, showed that the races, if they were to stay pure and thus survive, had to avoid mixing at all costs. It was a simple matter of species survival and empirical evidence. If he could only get Grant to go a little easier.[107]

Grant's book was becoming very popular among those committed to immigration restriction, Nordic supremacy, and other ideas of race and culture. It generated rave reviews from nativists and eugenicists around the country. Support for his work came primarily from amateurs obsessed with race and the breakdown of American order, and he was happy to have it. He was bewildered, however, that few professional scientists rallied to his cause. He seethed in paranoid anger that the scientific community either ignored or rejected him because they all 'seem to be either afraid of Boas or else impregnated with Socialism.'[108]

When American Museum ethnologist Robert Lowie (a Boas student) gave the book a less than flattering review in *The Freeman*, Grant sneered to Osborn that 'we should mark down Lowie as a very dangerous anthropological Bolshevist.'[109] He also wanted Osborn to find out if Lowie's name was originally Lowe. Grant became obsessed with Lowie and hated Franz Boas. Osborn took Grant's raving with a grain of salt. He contacted Lowie over the review, but refused to get agitated over it. He liked Lowie and respected his work. On the other hand, Osborn had an on-and-off relationship with Boas, who had worked at the museum from 1895 until 1905 before going to Columbia. Not only did he disagree with many of Boas's ideas, but Osborn's anti-Semitism was becoming more overt. He supported the editor of Henry Ford's anti-Semitic *Dearborn Independent* – who was then under suit for slander – and claimed that Franz Boas's preference for nurture over nature as the mechanism of social construction was due to Boas's 'Oriental affinities:' in other words, because he was Jewish.[110]

What makes Grant's anger at Lowie interesting is that Lowie was not opposed to eugenics or Nordic supremacy. 'I even believe that a fairly plausible case might be made out for racial inequality,' he told Osborn. His problem was with Grant personally. He admitted that reading the book 'convinced me that [Grant] is a charlatan,' and a pathetic joke.[111] Osborn confided to Lowie that he too felt Grant's work was overly biased and unscholarly. He told Grant that Lowie's critique was on target, and that Grant's book required 'rewriting from beginning to end.'[112] While he liked the book in general, Osborn was troubled with Grant's strident tone and lack of intellectual rigor (the first edition had almost no footnotes and a haphazard bibliography), and insisted in future editions that he bring the work up to higher standards.

Grant wanted his book read by the movers and shakers of American society. Among those Grant had Scribner's send copies to were Secretary of State Charles Evans Hughes and scientific 'colleagues' such as superintendent of the Eugenics Record Office, Charles Davenport, paleontologist William K. Gregory, archaeologist Flinders Petrie, and Theodore Roosevelt.[113] In addition, copies went to German eugenicists Eugen Fischer, Fritz Lenz, and future Nazi ideologist Alfred Rosenberg.[114] Henry Ford read and enjoyed the book.[115] Eventually, the journal *Science* favorably reviewed *The Passing of the Great Race*, calling it an 'interesting and valuable pioneer attempt at an interpretation of history in terms of race,' while the journal *Man* also spoke of its 'refreshing' tone and importance for spreading eugenics concepts to the general public.[116] The popular press also found good things to say. The *Boston Transcript* called *The Passing of the Great Race* 'a book to be read and considered by thinking men,' and said that 'Mr Grant utters a warning which should be heeded before it is too late,' while *Publisher's Weekly* called it 'vigorous and outspoken.'[117] The book received subtler reaction as well. Historian Allen Chase argues that George Horace Lorimer, as editor of the conservative and influential *Saturday Evening Post*, ran articles and editorials based on Grant's theories. There were also several works patterned very closely on *The Passing of the Great Race*, including Lothrop Stoddard's *Rising Tide of Color* and Charles W. Gould's *America, A Family Matter*, both of 1920. Immigration and eugenics had become a hot literary topic.

Restriction or Selection

Anti-immigrant feeling was nothing new in the early twentieth century. Immigration opponents were making their voices heard at least as early as the 1820s and 1830s. Samuel Morse, when he wasn't painting or developing the telegraph, was a virulent anti-Catholic and anti-immigrant activist who wrote numerous books on the subject, including *Imminent Dangers to the Free Institutions of the United States Through Foreign Immigration, and the Present State of Naturalization Laws* (1835).

The fight for control of New York's political power, and its relationship to the immigrant question, had continued to rage since 1894 when Madison Grant and the boys helped get William Strong installed into the mayor's residence at Gracie Mansion. By 1910 the Democratic Tammany Hall machine, still breathing after the Boss Tweed corruption and bribery scandals, took over again, only to be replaced by yet another Republican reform administration led by J.P. Mitchel. However, New York was growing into a union city and the Democrats regained power. By 1920, the Democratic Party was securely back in charge behind the leadership of the flamboyant Irish

Catholic mayor Jimmy Walker.[118] The new Democratic administration considered taking control of the zoological park away from Grant and the Society, putting it under city auspices.

This turn of events prompted Osborn to complain to Edward Poulton that the city now had a 'solid Irish Catholic government from top to bottom,' adding, 'you can imagine their attitude towards education and science.'[119] In addition to holding on to the park, he had been trying to get the city to underwrite the museum's expansion into a center of education for the public school system, but was running into resistance. Osborn felt that a special city education tax should be enacted to pay for educating the city's public school children, and that some of that money should go to the museum. The lack of education funds was due, he said, to the lopsided tax system which taxed only the wealthy and land- or home-owners, but not the poor, who paid no taxes, yet had children in the public school system (many of whom were not even citizens). Compounding the problem, Osborn complained that when a poor person did draw themselves up from their poverty, work hard, and save some money to buy a house or small plot of land, the government then taxed them so heavily that they were invariably ruined. This condition was causing 'discontent and Bolshevism' to spread, and had to be addressed by fairer, more equitable tax laws. Then, showing an astonishing lack of knowledge about tenement life, Osborn claimed that many tenement dwellers lived there so they would not have to pay taxes, and would save money on living expenses so that the excess could be sent back to the mother country instead of being spent in America where it would do the community good. Taxing everyone equally, Osborn argued, would provide the necessary funds to cover education costs and make available the money needed to expand the museum and make it a learning center.[120] He told New York governor Al Smith that 'aliens' were benefiting from public education but not contributing to the system's upkeep.[121] He did see hope, though. He boasted of a good rapport with Walker as well as the city comptroller, and the governor, who was a Catholic, but 'a very fine character.'[122] Osborn could talk of his Irish friends and how he admired them despite the fact they were Irish, then sarcastically complain that the city's newspapers would not run eugenics or immigration-restriction articles for fear of hindering 'circulation among the dear Irish.'[123]

Within this swirl of Republican activity, Osborn's younger brother, William Church Osborn, became a Democrat. After graduating from Princeton and receiving his law degree, Willie went into politics at the urging of his older brother. By 1913 he was a judge and close to Al Smith. Though a Democrat, Willie ran for office as an anti-Tammany Hall candidate like his grandfather, Jonathan Sturgis. His political power and aspirations growing, Willie ran for governor of New York in 1922, but lost to his friend Smith. Placing himself well as a politician of demonstrated honesty in New York, William Church Osborn became part of the intimate circle growing around the future president of the United States, but then Secretary of the Navy, Franklin Roosevelt, and worked closely with Woodrow Wilson. He married Alice Dodge, and his son Frederick Osborn went on to a career in eugenics and population studies, joining his uncle's Galton Society and American Eugenics Society, as well as serving as an American Museum trustee.[124] Politics was becoming integral to the eugenics movement. Theories meant little if no action could be taken. The point of focus for eugenicist political action came in the realm of immigration restriction.

Books were fine, but they were talk. Madison Grant wanted action, political action to combat the rising tide of what his friend Lothrop Stoddard called the undermen. The Immigration Restriction League had tried for years to make intelligence testing a prerequisite for new arrivals. He wanted more. Before 1900, eugenics in America had been a largely theoretical discussion. Afterwards, it became a political movement. Eugenicists around the country had moderate success with forced sterilization laws between 1907 and 1917. In 1905 the first major eugenics research facility, the Eugenics Record Office, opened at Cold Spring Harbor, Long Island, New York under the direction of biologist Charles Davenport, and with the financing from the Carnegie and Harriman fortunes.[125] Early sterilization laws were mostly shot down by local authorities by 1920, only to be revived after the Supreme Court upheld a state's right to forcefully sterilize a citizen in the *Buck* v. *Bell* case of 1927. Despite all this activity, the movement was fragmented and never really coalesced as a unified front.[126]

Other than the Quota Act of 1921, the Asian exclusion laws and scattered anti-miscegenation rules, the eugenics movement could not boast of a major political victory. It might also be argued that even these rulings were not really the result of direct action of eugenics promoters like Grant or Stoddard, but were due to local sentiment and prejudice. The one major political immigration victory where credit could be given the eugenicists was the Johnson/Reed Act of 1924. Albert Johnson, the Washington State Republican who gave his name to the act, was a close ally of Madison Grant, Henry Fairfield Osborn and other members of the New York conservationist circle.

Johnson was a rough-hewn, hard-drinking politician who had a strong dislike of immigrants, and staffed his House Committee on Immigration and Naturalization with eugenicist Harry Laughlin.[127] He also contributed a chapter to Grant's *The Alien in Our Midst* (1930) – a compilation of writings on immigration issues written by Grant, Osborn, Charles Davenport, Paul Popenoe, Harry Laughlin and others, as well as passages from the founding fathers. Grant called it a 'cross section of the best thought [on immigration restriction] of the United States.' In his section, Johnson boasted that 'I have been pegging away for restriction of immigration for more than thirty years.'[128] Allan Chase argues that the 'clique' of 'Davenport – Grant – Osborn' had Johnson installed as the president of the Eugenics Research Association in

1923 as a way of establishing manipulative power over him. Grant often had Johnson as a guest at his clubs and home, where they could plan strategy for implementing eugenics legislation.[129]

True to form, Osborn was uneasy with the way most eugenicists approached their subject. He was a professional scientist who insisted upon a professional bearing. Many eugenicists did not follow Osborn's precept about bearing and behavior, and often used crude rhetoric, name calling, and other lines of reasoning that he found impolitic. He did not like all the 'exploding, shrieking, pounding [and] raging' that characterized much of the movement's pronouncements.[130] He wanted a quieter, more reasoned tone. He even suggested that the Immigration Restriction League should change its name to Immigration Selection League as a way to make it a more respectable institution.[131] He was also wary of intelligence tests, arguing that they left loop-holes through which men of a 'Trotsky type' could wriggle through to spread their dangerous politics.[132] Despite his talk of moderation in discourse, he said little about actions. His preoccupation with evolution blinded him to the realities of his support for eugenics and racial superiority. He spoke out against allowing the eugenics movement to become a general race-hatred movement, but never spoke out against the specific violent acts like lynching, Black town burning, and violent anti-Semitism common in the 1920s and 1930s.

Grant had a hateful, even murderous intent towards immigrants. Osborn was more benign and paternalistic. Like his friend Theodore Roosevelt's view of native peoples (what Donna Haraway calls 'Teddy Bear Patriarchy'), Osborn thought the immigrants and poor were like children who could be good citizens, and who could contribute if they were properly taught.[133] They were not a lost cause; they too could be redeemed. Osborn saw the American Museum as the perfect vehicle for turning the dangerous hordes into productive citizens who knew their place in the overall scheme. In the museum and at the zoological park people could learn that 'nature is the visible expression of the divine order of things, and [that] her facts are immutable.'[134] The study of nature was 'the best means of raising the mental and moral level of our foreign-born population.' Though child-like, immigrants had the right to vote and affect politics. Since these child–adults in their large numbers could and did sway elections, it was the responsibility of community leaders to teach them how to understand democracy and appreciate the greatness of their adopted country. Like children they would learn simple lessons the best. He devised a mnemonic ditty to teach the newcomers the hierarchy of nature and how they fit into it:

> Man is part of nature
> Nature is the act of God.
> Through obedience to nature,
> Man obeys the will of God.[135]

As the 1920s progressed, Osborn's views hardened, but he exhibited the same type of ambivalence he showed towards the new biology. He sympathized with working-class immigrants and small shop owners who he said were being taken advantage of by wealthy Jewish and Italian landlords, but in the same breath could then accuse those same Italian grocers and small business men of being 'unscrupulous' and using business methods 'which the Americans will not do.' He claimed that the Italian grocers in his Garrisons enclave were engaged in the illicit whiskey trade and mixed up in organized crime. But he had to admit that the Italians and the Jews were hard workers. 'I like the Italians,' he said, and even 'the Japanese have many virtues.'[136]

When the Johnson Act passed in 1924, Osborn told its namesake that it was one of the most important steps taken 'in the whole history of our country.'[137] As good as it was, the Johnson Act did nothing about race mixing. Osborn had seen in the fossil record how unrestricted immigration and race mixing had wrecked past species. Not only Nordics, but all races would suffer from miscegenation. If the races mixed, it could throw off the entire scheme of evolution, upset the special attributes they had. The separate races should be preserved intact. In his introduction to *The Passing of the Great Race*, Osborn insisted that racial preservation 'is not a matter either of racial pride or racial prejudice.'[138] It was something meant to improve the lot of the country and thus the lot of all its people. He admitted that, though they were not on the same level as Nordics, even the lowliest races had a certain divine spark of genius which should be cultivated. What Osborn wanted was 'not more people but better people, not more Americans but better Americans.'[139] If Osborn's views on human hierarchies were to be supported by the fossil record, as he argued they could be, he would have to find some, and find some in Central Asia especially.

Notes

1. HFO to Josephine Osborn, 6/7/1908, box VIII, OSNY.
2. HFO, box XI [1915?], OSNY.
3. Quoted in Stephen Jay Gould, *The Mismeasure of Man* (New York: W.W. Norton & Co., Inc., 1996), 261.
4. Richard Hofstader, *The Age of Reform* (New York: Vintage Press, 1955), 177.
5. John Higham, 'A Brake on Nativism,' in *The Progressive Era*, ed. Arthur Mann (New York: Rinehart & Winston, 1963), 48.
6. Irving Howe, *World of Our Fathers* (New York: Harcourt, Brace, Jovanovich, 1976).
7. Brian J. Cudahy, *Under the Sidewalks of New York: The Story of the Greatest Subway System in the World* (New York: Fordham University Press, 1995).
8. Robert P. Swierenga, 'Religion and Immigration Behavior: the Dutch Experience,' in *Belief and Behavior: Essays in the New Religious History*, eds Philip

R. Vandermeer and Robert P. Swierenga (New Brunswick: Rutgers University Press, 1991), 165–188.

9. Samuel P. Hayes, *The Response to Industrialism: 1885–1914* (Chicago: University of Chicago Press, 1957).
10. Letters between HFO and Jacob Schiff, March 1912, box VIII, OSNY.
11. Robert Wiebe, *The Search for Order: 1877–1920* (New York: Hill and Wang, 1980), 166.
12. John Radford, 'Sterilization Versus Segregation: Control of the "Feebleminded" 1900–1938,' *Society of Science and Medicine*, 33:4 (1991), 449–458.
13. Mark Haller, *Eugenics: Hereditary Attitudes in American Thought* (New Brunswick: Rutgers University Press, 1963).
14. For the eugenics movement see Diane Paul, *Controlling Human Heredity* (Atlantic Highlands, NJ: Humanities Press, 1995), 103, and *The Politics of Heredity* (Albany: State University of New York Press, 1998), 105. Also Daniel Kevles, *In the Name of Eugenics* (Cambridge: Harvard University Press, 1995), 75.
15. HFO, *Creative Education* (New York: Charles Scribner's Sons, 1927), 306.
16. Edward Drinker Cope, 'The Marriage Problem,' *The Open Court*, 64 (November 15, 1888), 1307–1324.
17. Edward Drinker Cope, 'The Relation of the Sexes to Government,' *Popular Science Monthly* (October, 1888), 730. Also see Susan Sleeth Mosedale, 'Science Corrupted: Victorian Biologists Consider "The Woman Question,"' *Journal of the History of Biology*, 11:1 (Spring 1978), 1–55.
18. Edward Drinker Cope, 'Two Perils of the Indo-European,' *The Open Court*, 3: 126 (January 23, 1890), 2052–2071.
19. HFO, *Cope: Master Naturalist*, 571.
20. Charles C. Alexander, 'Prophet of American Racism: Madison Grant and the Nordic Myth,' *Phylon*, 23:1 (1962), 73–90. John Higham, *Strangers in the Land* (New York: Athenaeum, 1973), 272. Allan Chase, *The Legacy of Malthus* (New York: Alfred A. Knopf, 1977), 173. Kevles, *In the Name of Eugenics*, 75. Carl N. Degler, *In Search of Human Nature* (New York: Oxford University Press, 1991), 48–51. Paul, *Controlling Human Heredity*, 103. Stephan Kuhl, *The Nazi Connection* (New York: Oxford Press, 1994), and Robert Proctor, *Racial Hygiene: Medicine Under the Nazis* (Cambridge: Harvard University Press, 1988) deal with Grant as an inspiration to German eugenicists.
21. Gabriel Grant biographical file, New Jersey Historical Society. Gabriel may have taken his son's name from either the location of the hospital or the boy's place of birth, Madison Avenue, New York.
22. Gabriel Grant Papers, MG251, New Jersey Historical Society.
23. Madison Grant to Maxwell Perkins, 2/15/1922, box 67, Author Files I, Scribner Archive, PU. Grant had a sister who died in 1909 as well as two younger brothers: DeForest (who, like Madison, never married) and Norman. This third Grant son married Sophie Cole in 1906, and apparently died early, since he is not mentioned in Madison's obituary as surviving. Gabriel Grant obituary, *Proceedings of the New Jersey Historical Society*, 3: 10 (1915), 87–89. Madison Grant obituary, *Proceedings of the New Jersey Historical Society*, 55 (1937), 302.
24. *New York Times*, obituary (May 31, 1937).
25. Madison Grant to HFO, 5/3/1917, CAAM.
26. Madison Grant to HFO, 2/27/1929, CAAM.
27. He did produce a series of wildlife monographs including 'The Caribou'

(1902), 'The Moose' (1903), 'Origins and relationships of the Large Mammals of North America' (1904), and 'Saving the Redwoods' (1919).

28. The Dresden naturalist, Arnold Jacobi, told a curator at the American Museum that this animal did not deserve to be treated as a separate species. 'Rangifer Granti ... has been named on such worthless characters that not even the significance of subspecific rank can be accorded it.' He did like Grant, however. Arnold Jacobi to AMNH, 11/1/1932, MAAM.
29. HFO to Fairfield Osborn Jr, 1/12/1921, box IX, OSNY.
30. Quoted in Jonathan Spiro, *Patrician Racist: The Evolution of Madison Grant* (University of California at Berkeley: Ph.D. diss., 2000).
31. John Foord, *The Life and Public Service of Andrew Haswell Green* (Garden City, New York: Doubleday Publishing, 1913).
32. HFO to Fairfield Osborn Jr, 1/12/1921, box IX, OSNY.
33. HFO, typescript copy of article on the Park's history, box XXIII, OSNY.
34. *New York Times*, obituary (May 31, 1937). Grant's other building project was the Taconic State Parkway. He served as the Zoological Society's secretary, chairman of the executive committee, and president until his death. For reform politics see Samuel Hays, 'The Politics of Reform in Municipal Government in the Progressive Era,' *Pacific Northwest Quarterly*, lv (October 1964), 157–169.
35. Theodore Roosevelt, 'Race Decadence,' *Outlook*, 97 (1911), 763–769.
36. John Reiger, *Passing of the Great West: Selected Papers of George Bird Grinell* (New York: Scribner's, 1972).
37. Madison Grant, 'History of the Zoological Society,' *Zoological Society Bulletin*, 37 (January, 1910), and 'The Origin of the New York Zoological Society,' in George Bird Grinell and Theodore Roosevelt, eds, *Trail and Campfire: The Book of the Boone and Crockett Club* (New York: 1897). Also see William Bridges, *Gathering of Animals: An Unconventional History of the New York Zoological Society* (New York: Harper & Row, 1974).
38. William Temple Hornaday, *Our Vanishing Wildlife: Its Extermination and Preservation* (New York: New York Zoological Society, 1913).
39. Helen Horowitz, 'Animal and Man in the New York Zoological Park,' *New York History*, 55 (1975), 426–453.
40. *New York Sun* (March 15 & 16, 1900).
41. HFO to Madison Grant, 8/21/1928, CAAM.
42. Various letters in Grant Papers, 1902–36, MAAM. For more on conservatism, the eugenics movement and conservation see Gray Brechin, 'Conserving the Race: Natural Aristocracies, Eugenics, and the U.S. Conservation Movement,' *Antipode*, 28:3 (1996), 229–245.
43. *New York Herald* (6/2/37).
44. E.W. Nelson to Madison Grant, 5/1933, MAAM.
45. Madison Grant to Phillips, 6/30/32, MAAM.
46. Letters between Grant and Senator Albert Johnson, 1932, MAAM.
47. Barbara Solomon, *Ancestors and Immigrants* (Cambridge: Harvard University Press, 1956), 101–102, 201–202.
48. HFO to Lucretia Osborn, 3/20/1902, box VII, OSNY.
49. HFO, 'Opening of the Hall of Public Health,' in *Articles on J.P. Morgan and the American Museum* (New York: *American Museum Journal*, 1913), 195.
50. HFO, 'The American Museum and Citizenship,' 54th Annual Report of the Trustees of the American Museum of Natural History for the year 1922 (1923).
51. HFO, 'The Fighting Ability of the Different Races,' *Journal of Heredity*, 10:1 (1919), 29–31.

52. HFO to William Berryman Scott, 8/3/1914, box IX, OSNY.
53. HFO to Edward Poulton, 10/14/1914, box IX, OSNY.
54. HFO to Edward Poulton, 6/16/1915, box IX, OSNY.
55. H.L. Mencken, 'Theodore Roosevelt, Psuedo Liberal,' in *A Mencken Chrestomathy* (New York: Alfred A. Knopf, 1920), 76.
56. HFO to Edward Poulton, 1/27/1917, box IX, OSNY.
57. Ibid.
58. HFO to Henry Warren Osborn, 6/11/1918, box IX, OSNY.
59. See Spiro, *Patrician Racist.*
60. Madison Grant to Charles Scribner, 6/22/1916, box 67, Author files I, Scribner Archive, PU.
61. Madison Grant, *The Passing of the Great Race* (New York: Charles Scribner's Sons, 1916), 80.
62. Ibid., vii–ix.
63. For more on objectivity and racial questions see Halford Fairchild, 'Scientific Racism: the Cloak of Objectivity,' *Journal of Social Issues*, 47:3 (1991), 101–115.
64. Oliver Allen, *New York, New York* (New York: Macmillan Publishing Co., 1990).
65. HFO to Edward Poulton, 5/31/1920, box IX, OSNY.
66. HFO to Edward Poulton, 5/31/1920, CAAM, and *New York Times* (7/2/1922).
67. HFO to Josephine Osborn, 6/6/1919, box IX, OSNY.
68. HFO to Josephine Osborn, 5/18/1919, box IX, OSNY.
69. Josephine Osborn to Lucretia Osborn, 1/11/1922, box IX, OSNY.
70. HFO to Josephine Osborn, 7/29/1926, box X, OSNY.
71. See Perry Osborn to HFO, 6/5/1923, and Henry Fairfield Osborn Jr to HFO, 2/29/1924, CAAM.
72. HFO to Carrie Chapman Catt, 12/28/1917, CAAM.
73. Dororthy Parker, 'Arrangement in Black and White,' *The Portable Dorothy Parker* (New York: Penguin Books, 1976), 20.
74. Ann Douglas, *Terrible Honesty: Mongrel Manhattan in the 1920s* (New York: Noonday Press, 1995).
75. HFO, 'Address of Welcome,' Second International Congress of Eugenics, Eugenics and the Family, I (Baltimore: Williams & Wilkins Company, 1923), 2.
76. *New York Times* (11/29/1925).
77. HFO, 'Selecting the American Child,' *Kindergarten and First Grade Magazine*, 10 (September 1924), 5–7.
78. *The Peking Leader* (10/10/1923).
79. HFO, *Impressions of the Great Naturalists* (New York: Charles Scribner's Sons, 1924).
80. HFO interview with Walter Tittle, *The World's Work* (May 1928).
81. HFO to James R. Angell, 7/2/1929, Davenport Papers, American Philosophical Society. Hereafter APS.
82. Michael Ruse, *Monad to Man* (Cambridge: Harvard University Press, 1996), 268.
83. Madison Grant to William K. Gregory, 4/18/1918, Davenport Papers, Galton Society Folder, APS. For his part, though he certainly supported all the society's aims, Osborn seems also to have used it as a way of making contacts for acquiring fossil human casts for his work; in particular he wanted a set of Eugene Dubois's Java Man remains.

84. HFO to A.J. Balfour, 5/14/1917, box IX, OSNY.
85. HFO to Stewart Paton, 5/10/1921, CAAM.
86. Letters between HFO and Charles Davenport, 1914, Davenport Papers, APS, and HFO Diaries, box XXV, OSNY.
87. Mark Haller, *Eugenics: Hereditarian Attitudes in American Thought* (New Brunswick: Rutgers University Press, 1963), 73–74.
88. Charles Davenport to HFO, 5/8/1920, Davenport Papers, APS.
89. Herbert Spencer Jennings to HFO, 6/7/1920, Davenport Papers, APS.
90. HFO to Charles Davenport, 10/19/1920, Davenport Papers, APS.
91. HFO, 'Address of Welcome.'
92. See *New York Times* (11/21/1921), and HFO to Charles Davenport, 11/26/1921, CAAM.
93. HFO to Irving Fisher, 6/21/1922, CAAM.
94. HFO, 'Lo, The Poor Nordic,' *New York Times* (4/8/1924).
95. Ibid.
96. *New York Times* (4/13/1924).
97. Walter Lippmann to HFO, 4/22/1924, CAAM.
98. HFO to Walter Lippman, 4/29/1924, CAAM.
99. H.B. Hayden to HFO, 4/8/1924, and HFO to Charles Davenport, 8/21/1924, Raymond Pearl Papers, Galton Society Folder, APS.
100. HFO to Leon Whitney, 6/8/1927, CAAM.
101. HFO, 'Osborn Calls US Europe's Dump,' *New York Sun and Globe* (12/13/1923), 1–2.
102. He used the word emigrant instead of immigrant.
103. HFO, 'Race Progress in its Relation to Social Progress,' *Journal of the National Institute of Social Science*, ix (November 10, 1924), 8–18.
104. HFO to Madison Grant, 12/23/1919, CAAM.
105. HFO to Madison Grant, 12/19/1922, CAAM.
106. HFO, 'The Cavemen Knew,' *Collier's Magazine*, 75:21 (May 23, 1925), 23.
107. *New York Times* (9/25/1921).
108. Madison Grant to Charles Davenport, 2/16/1917, Grant folder, Davenport Papers, APS.
109. Madison Grant to HFO, 6/5/1922, CAAM.
110. HFO to Delancy Nicoll, 5/22/1926, CAAM.
111. Robert Lowie to HFO, 6/19/1922, CAAM.
112. HFO to Madison Grant, 6/6/1922, CAAM.
113. Though Madison Grant initially worked with Charles Scribner on *The Passing of the Great Race*, by 1918 he was working with editor Maxwell Perkins who also edited the work of F. Scott Fitzgerald and Ernest Hemingway.
114. Grant gave lists to Scribner's that often had as many as a dozen or more names on them. He wanted so many complimentary copies that Scribner's eventually told him to pare the lists down or pay for them himself. Lists in Grant Papers, Scribner Archive. PU.
115. Henry Ford to Charles Scribner, 10/28/1921, box 67, Author files I, Scribner Archive, PU.
116. H.J.E. Peake, 'Review of *The Passing of The Great Race*,' *Man*, 20:173 (November, 1920).
117. *Boston Transcript* and *Publisher's Weekly* (9/12/1916).
118. Allen, *New York, New York*.
119. HFO to Edward Poulton, 5/31/1920, box IX, OSNY.
120. *New York Times* (5/28/1919).

121. HFO to Otto Kahn, 8/16/1920, and other correspondence in the Otto Kahn Papers, PU.
122. HFO to Edward Poulton, 5/31/1920, box IX, OSNY.
123. HFO to Madison Grant, 6/21/1920, CAAM.
124. For William Osborn's political career see correspondence in box 6, Osborn/Dodge Papers, PU. For Frederick Osborn's career see Harry Shapiro, 'In Memory of Frederick H. Osborn,' *Social Biology*, 29 (1982), i–iii.
125. For a history of the ERO and Grant's relationship to it see Garland Allen, 'The Eugenics Record Office at Cold Spring Harbor, 1910–1940, An Essay in Institutional History,' *Osiris*, 2, 2nd edn (1986), 225–264.
126. Gould, *Mismeasure of Man*.
127. Kevles, *In the Name of Eugenics*, 103.
128. Madison Grant and Charles Stewart Davison, *The Alien in Our Midst* (New York: Galton Publishing Co. Inc., 1930), 30.
129. Chase, *Legacy of Malthus*, 289, 295.
130. HFO to Madison Grant, 5/28/1920, CAAM.
131. HFO to Madison Grant, 12/12/1922, CAAM.
132. HFO to Madison Grant, 12/13/1922, CAAM.
133. Donna Haraway, *Primate Visions* (New York: Routledge, 1989), 26.
134. HFO, 'The American Museum and Citizenship,' 54th Annual Report of the Trustees of the American Museum of Natural History for the year 1922 (1923), 1.
135. Ibid., 17.
136. HFO to Charles Evens Hughes, 3/25/1924, and HFO to Madison Grant, 6/27/1928, CAAM.
137. HFO to Albert Johnson, 5/2/1924, CAAM.
138. Grant, *Passing of the Great Race*, ix.
139. HFO to James R. Angell, 7/2/1929, Davenport Papers, Osborn folder, APS.

CHAPTER SIX

'Go and Find Them'

For Osborn the Nordics were superior for the same reason any superior animal group was: they possessed a superior race plasm, followed the natural laws of evolution, and had enhanced their greatness by struggling to go further. In addition, the Aryans were a more generalized form, closer to the original point of human radiation. This allowed them to adapt more easily to changing conditions and overcome them. Non-Aryans had become biologically specialized to fit their particular niches, and thus were less able to advance. Africans, Asians, Australian Aborigines and other people of color occupied the far-flung edges of the world while Europeans were centrally located. As he believed it did for the other mammals, Osborn believed this geographic distribution, and level of biological specialization, showed Central Asia to be the home of the human race.

The final step to solidifying this complex theory was to get the physical proof in the form of fossils from Central Asia. Unfortunately, Osborn was not a hardcore explorer or field man. Though he railed against the experimental biologists for spending too much time in the laboratory and not in the field, he himself was never really comfortable in the wilds and was never more than a tourist at the sites his people worked. Had it depended on Osborn going to Central Asia himself, it would never have happened. He had too much on his hands running the museum and writing his various books and articles. He had convinced himself of the Central Asia hypothesis: acquiring physical proof would go a long way to proving it to others. For philosophers like Hegel and Haeckel, and the theosophist Blavatsky, Central Asia would remain a dream – a romantic icon, never approached, never touched, safely out of reach where it could never be disproved. It would have been the same for Osborn had it not been for Roy Chapman Andrews.

The Central Asiatic Expeditions

Raised in the wilds of Wisconsin, Roy Chapman Andrews (1884–1960) was a natural explorer, boasting that 'I happen to have been born to it.'[1] He attended preparatory school and then Beloit College. In 1906 he went to New York City with thirty dollars in his pocket and a burning desire to work for the American Museum of Natural History. He was hired by museum director H.C. Bumpus as an assistant in the taxidermy department, where his duties included scrubbing the floors and doing other menial work. This lowly position did not discourage his ambition and he soon worked his way up to more significant assignments.

Eager to do anything, Andrews was sent on a number of missions to collect whales on Long Island. Indeed, for a time it seemed as if Andrews's life work would be the study of whales. His leadership abilities were shown when he was sent on a whale-collecting expedition to Asia in 1909. He traveled through the Dutch East Indies, Japan, Korea and China. Continuing west through Europe, he returned to New York in 1913 with a reputation as a dauntless explorer. That same year he married Yvette Borup, the sister of an Arctic explorer. A kindred soul, Yvette accompanied her husband on most of his travels and eventually produced two sons. At the same time he completed a Master's degree in mammology at Columbia University, where he worked with William K. Gregory and attended the last classes of Henry Fairfield Osborn. Andrews was in awe of Osborn and would do anything to work with him. In his early days as a floor washer, Andrews would hide behind the meteorites in the museum just to watch the great man walk by.[2]

As much fun as it was, Andrews did not want to spend the rest of his life chasing whales. As early as 1912 he began formulating plans for a major Asian undertaking. He had become enchanted with Asia and entranced by Osborn's 1900 paper which pointed to Asia as the cradle of origin for mammals and humans. Andrews longed to prove the theory correct. He approached Osborn with a proposal to do some reconnaissance missions to Asia in 1915. He did not tell Osborn the full extent of his idea at the time, however. Osborn agreed to some limited work, and Andrews set off for China to check out routes and do a feasibility study for a monumental exploration of Central Asia. With the outbreak of the Great War Andrews's grandiose plans were put on hold while he worked for US Naval Intelligence in China. Just what he did as a spy is unclear. Even as late as 1929 he would only say: 'I shall have to omit a record of that part of my life.'[3]

With the war over, Andrews could get back to business. In January 1920 he had lunch with Henry Fairfield Osborn at the museum. After some small talk Osborn asked him what was on his mind. Andrews replied: 'I want to test your theory. I want to see if I can find primitive human remains or traces of man's evolution.' Osborn sat and listened transfixed as Andrews laid out his plans for the expedition. Here was a man who could go to Central Asia and find the very evidence he needed to turn the theory into fact. He saw in Andrews the embodiment of the explorer ideal: intelligent, resourceful, fearless and possessed of just the sort of indomitable will-power Osborn believed was necessary to overcome predestination. Andrews said the massive expedition would succeed because he 'wanted to do it so intensely that [he] was willing to make any sacrifice.'[4]

Andrews laid out for Osborn a multifaceted, interdisciplinary scientific survey of Central Asia. He wanted to bring experts in different fields to study not only the geology, but the botany, paleontology, climatology, archaeology and history of the region. Nothing of the kind had ever been proposed. He would use cars and trucks instead of the traditional camels to cover huge areas. In this way he could learn more about Central Asia in a few years than had been learned in centuries of previous exploration. When Andrews finished his presentation he slumped back in his chair, happy to have unburdened himself, but apprehensive about how Osborn would react. He need not have worried. Osborn was all smiles. He beamed at Andrews like a proud father, just as he had beamed on Fairfield Jr's eighteenth birthday. All he said was, 'Roy, we've got to do it.'

After talking with Osborn, Andrews went to the library of J.P. Morgan and pitched him on the idea. Morgan listened intently and then wrote out a check for fifty thousand dollars as his official contribution, and a second fifty thousand on the side. Andrews then spent a grueling year lunching, dining and lecturing New York's robber barons to raise more money. He received donations from philanthropists such as John D. Rockefeller Jr, Childs Frick, president of the First National Bank, George Baker, Sidney Colgate and E.H. Gray of US Steel.[5]

Both Andrews and Osborn thought it would be a good idea to announce the expeditions publicly.[6] The fact that they discussed the search for the origins of humankind struck a responsive chord amongst the public and caused a great stir in the media. They would later regret making the search for a missing link such a prominent part of the press releases. Almost immediately after announcement of the expeditions, the museum was flooded with applications from people wanting to go along. Andrews, of course, had already picked most of the team from the museum staff. He also planned to hire local help in China and Mongolia in the form of camp technicians, cooks, teamsters and interpreters. It was cheaper and smarter to have local people who knew the language and the terrain, and who had experience of the harsh field conditions of Central Asia.

The letters that poured into the museum from people wanting to go on the expedition ran the gamut from the reasonable to the laughable to the bizarre. While several women said they wanted to accompany Andrews to keep him from getting lonely, one telegraphed on the day of the press conference offering help of an otherworldly type. 'Regarding search for missing link,' she said, 'Ouija board offers assistance.' From Andrews's account of this individual (he never gave her name) she seems to have been at least familiar with the work of Madame Blavatsky and theosophical belief. Possibly referring to Madame Blavatsky's Shamballa, she said that Andrews would find a secret buried city in the desert. In this city he would find the history of man from 'the time he crawled on all fours until the dawn of history.' She learned this, she said, by consulting 'certain spirits.' Not one to pass up an opportunity, Andrews wrote back to her asking for more precise longitudes and latitudes for finding the lost city. Sensing that Andrews was not taking her advice seriously, she replied that the spirits were annoyed at his flippant attitude, and he should just look for four large stones buried in the sand.[7] There is no evidence that Andrews was interested in the occult. He apparently missed what may have been theosophical references in the woman's letters. It makes one wonder, though, how an expedition like this must have stirred the souls of theosophists and occultists now that someone was actually going to explore for the origins of man in the Gobi Desert to confirm what they already knew. In their minds he could very well stumble upon the Hidden Masters and their Gobi fortress, just as Madame Blavatsky had done years before.

As media attention grew, Andrews and Osborn became concerned over the public focus on the missing link.[8] The two scientists were afraid such a preoccupation might taint the expedition in the eyes of the scientific establishment. Fossil men were only a part, albeit a major part, of a much larger agenda of working out the natural history of the region. Andrews tried desperately to explain all this, but to no avail. Finally, seeing the inherent publicity problems, the search for early man was quickly and quietly dropped from most of the official pronouncements about the expedition and its goals. The missing-link aspect had grabbed the public's attention, however, and they would never let go of it.

The pestering came from inside as well as outside the museum. Trustee Madison Grant, who was also on the board of the New York Zoological Society, cornered Andrews one day to ask a favor. In his early days at the museum Andrews had lectured to various groups on Grant's behalf and so the two were familiar with each other. Grant wanted Andrews to bring back a live panda for the zoo's collection. Andrews balked at the idea. Bringing back a skin or skeleton was one thing, but a live animal was another. Besides, the United States government had put up an embargo against bringing in alien animals that might bring disease with them. Grant insisted that Andrews take the $2,000.00 he was offering to cover expenses. Finally, Andrews told Grant he would see what he could do, but not to count on it.[9] Andrews had far more important things to worry about.

The logistics of the Third Asiatic Expedition were enormous, but the attack plan was relatively simple. Andrews had already sketched out a rough trail for the expedition to follow. From their headquarters in Peking (Beijing), China, the automobile convoy would head out across the Mongolian frontier stopping at localities where the geology suggested fossils might be found. They would prospect briefly before heading for the next location. Meanwhile, huge lumbering camel caravans, loaded with fuel and other provisions, would meet them at regular intervals to drop off supplies and take in fossils. In this fashion they could cover enormous areas quickly and efficiently. At the end of the summer they would head back to Peking to winter over, examine their finds and prepare for the next field season.[10]

Fig. 7 Osborn, truck and horses in Mongolia, 1925. Courtesy Department of Library Services, American Museum of Natural History.

In February 1921, with the money procured, a team assembled and equipment gathered, Andrews and the expedition left New York. Osborn did not go, but saw them off at the dock. At the last minute Andrews expressed some of his rare doubts about the whole enterprise. Osborn shook his hand, clapped him on the shoulder and said, 'The fossils are there, I know they are. Go and find them.'[11]

Upon arriving in China they ensconced themselves in a villa rented from a Peking aristocrat.[12] The Third Asiatic Expedition left Peking watched by a crowd of curious, skeptical Chinese. The heavily laden cars trundled precariously through the Great Wall at Kalgan and out into the Mongolian Desert. Travelling with them briefly was Davidson Black, a young Canadian anatomist from the local Peking Union Medical Hospital, who was looking to collect some anthropological data for his own search for the earliest human fossils.

Andrews and Osborn had been told that it was pointless to search for fossils in the Gobi Desert. Wags said that they might as well look for traces of ancient humans at the bottom of the sea. It has been noted that before the Central Asiatic Expeditions the only fossil ever found in Central Asia was a single rhinoceros tooth picked up in the 1890s by a Russian explorer named Obruchev.[13] It would seem a thin reed on which to hang such a costly undertaking by Osborn. It is easier to understand when it is considered that Osborn built his Central Asia hypothesis, and his adaptive radiation theory, on the history of the horses and titanotheres. The nearest relatives to the horses and the titanotheres are the tapirs and the rhinos. If fossil rhinos were in the Gobi, then it was likely that fossil horses and titanotheres had been as well. If they were, then it was likely that Central Asia was the point of mammal origins. So while it was a gamble to go to the Gobi Desert with only a single fossil tooth as evidence, it did fit the theoretical construct Osborn was working with. Richard Swann Lull, the Yale paleontologist, thought Osborn on the right track. Lull had been a student of Osborn's and in a lecture delivered at Yale in 1921 he backed his old teacher's assertions. He said that it had been 'seemingly established' that Asia was the cradle of man.[14] Still, not everyone thought Osborn was on the right track. As early as 1891, Smithsonian Institution scientists were saying that 'on any view of probabilities it is in the highest degree unlikely that we shall ever discover the exact cradle of our race.'[15] Nor were they so sure about the home of the Aryans either. One claimed that 'the highlands of Central Asia are no longer accepted without question as the cradle of the Aryan family.'[16]

It did not take long to silence the nay-sayers. The caravan's first major destination was the Mongolian capitol of Urga. On their way there the explorers planned to camp at the telegraph station of Iren Dabasu, nothing more than a cluster of mud hovels grouped like frightened children clinging together for warmth on the Gobi Plain. Andrews sent team paleontologist and second-incommand Walter Granger to scout for fossils while he set up camp. As Andrews was finishing pitching the tents, Granger's car came roaring in. Granger and the two geologists, Charles Berkey and Frederick Morris, jumped out, each with an armfull of reptile fossils. Barely able to control his excitement, Granger said, 'The stuff is here!'[17] Not only was the stuff there, but so were other explorers.

Though it was the largest, the Third Asiatic Expedition was not the first nor the only Western expedition in Central Asia. Explorers had been coming there, and continued to come, for years. Usually going alone or in pairs, adventurous men and women were roaming the wastes of Central Asia even as Andrews and his team arrived. Some came for the sense of discovery, like the Swede Sven Hedin; some came for spiritual and occult discovery, like Frenchwoman Alexandra David-Neel; some to spread Christianity, like Mildred Cable and Francesca French,[18] and some to spread empire like, the Briton Frances Younghusband. In fact, during the early decades of the twentieth century, Central Asia virtually spilled over with Western explorers.[19] While none of these individuals was looking for the origins of the human race, some were looking for answers as cosmic as Henry Fairfield Osborn was. Alexandra David-Neel was an actress turned occult adept who, having studied the work of Madame Blavatsky and other theosophists, was searching Tibet for philosophical secrets.[20] Nicholas Roerich, a Russian-born American, was also searching for mystical answers. He too studied theosophy and believed he could find Madame Blavatsky's fabled city of Shamballa: the same city Roy Chapman Andrews's correspondent told him to look for.[21] Andrews and his team were far from alone.

Mongolia

By the 1920s Roy Chapman Andrews had lived almost continuously in Asia since 1909 and spoke most of or parts of many of the local languages and had grown to love Asian culture. He seems to have had a genuine admiration for the Mongolian people, and was sympathetic to their plight of being caught between the Russians to the North and the Chinese to the East. They reminded him of colonial Americans in their determination to live independent, self-sufficient, honorable lives. He chronicled their culture, habits and society in his monumental recapitulation of the entire decade-long expedition, *The New Conquest of Central Asia* (1929). The Chinese, on the other hand, did not fare well in his eyes. Though he found many individual Chinese worthy of praise, he found them lacking as a group. Chinese society seemed to him one endless round of payoffs and kickbacks, 'squeeze' as it was called. Nothing would get done without plenty of 'squeeze' to grease the wheels of progress.

Fig. 8 Roy Chapman Andrews – the man with the plan. He could laugh his way through a hail of gunfire, then get teary-eyed over the fate of the Mongols. He once attacked a sandstorm with a pistol.

Neg. No. 258467, courtesy Department of Library Services, American Museum of Natural History.

The admiration Andrews felt for the Mongols was tempered by his fear that they were not long for existence. The Mongol 'race' was in decline, he said, due to the introduction of what he called 'Lamaism.' This religion, with what he considered its priest-class and anti-war, anti-violence, anti-ambition theology, had reduced them from a 'proud warrior race' to a groveling shell of their former greatness. As far as he was concerned this was all the fault of Chinese leaders who had systematically introduced the religion as a way of subjugating the fiercely independent Mongols. Where they were once 'virile and glorious' they were now desolute and slothful.[22]

Henry Fairfield Osborn arrived in China in 1922. Loulu came with him, but she stayed safely at the Peking headquarters with Yvette Andrews while the boys went out to play in the Mongolian mud. In his usual style Osborn came only to look and have photos taken. Several pictures show him sitting snugly in the back of a car while the rest of the team appear tired and dirty. Like Andrews, Osborn admired the Mongolians, but not the Chinese. He saw middle-class Chinese as 'callous, self-indulgent, effeminate [and] always sleeping.' He did have to admit, though, that there were intelligent Chinese and that the coolies had good physiques and senses of humor.[23]

China had suffered a series of political upheavals in the early part of the century. By 1912 it was enduring the period of the Warlords, where military strongmen held control of different parts of the country and fought with each other for control. Meanwhile the imperial government had descended into ineffective chaos. Nationalist leader Sun Yat Sen tried to create a Chinese republic run by the parliamentary Goumindong Party. With this the last emperor, Pu Yi, was forced to abdicate. Andrews and his wife attended the hapless Pu Yi's wedding. Amidst the splendor of the event they could not help but feel sad that they were witnessing the end of one of the great empires in human history. Marxism and revolutionary politics filtered into China from Russia and, in 1921, as the Third Asiatic Expedition was arriving, the Chinese Communist Party was formed. Though the Communists and the Goumindong at first attempted to work together for the betterment of China, they soon came to blows, and by 1927 civil war broke out, sending the country deeper into anarchy.

All this had consequences for the Third Asiatic Expedition. Andrews had to cross China to enter Mongolia, and he needed China as a base of operations. In Mongolia the political climate was not much better. The country was the rope being used in a tug of war between various parties. An army of renegade White Russians, forced out after the October Revolution, and led by a disgruntled baron named Ungern-Steinberg, were on a crusade to force out the Chinese. In their frenzy they slaughtered not only Chinese and Bolsheviks, but also Jews and Mongols until 1921, when they were annihilated by a Russian army sent to hunt them down. A Soviet-aligned government was installed and some sense of order restored. The name of the capitol city,

Fig. 9 Osborn in pith helmet in China, 1925. Courtesy of the Department of Library Services, American Museum of Natural History.

Urga, was changed to Ulan Bator (Red City). Andrews lamented the Soviet takeover of the prideful Mongols, and predicted their end because of it. 'The establishment of a Bolshevik regime is,' he said, 'the last act in the tragedy of Mongolia.' He finished dramatically by saying, 'the doom of the Mongols, as a race, is sealed.'[24]

Besides Andrews there were others in Asia looking for the remains of fossil humans. Davidson Black (1884–1934) was briefly associated with Andrews and his team in 1921. Black became interested in human origins while he was studying comparative anatomy with the famous British anatomist Grafton Elliot Smith (one of the Piltdown doubters) in 1914. When offered a position with the Peking Union Medical School he accepted, in part because he wanted to search for human fossils in Asia. Like Osborn, Black firmly believed in the Central Asian origins of man. When Andrews and his missing-link circus rolled into China, Black introduced himself and Andrews offered to take him along. Black only stayed with the team a short time: his school preferred he keep to his medical duties. In 1926 a pair of human fossil teeth were discovered outside Peking at Zhoukoudian. Black obtained a Rockefeller grant to continue work in 1927 and found a third tooth. By 1929 more material, including an almost complete skull, was found. Author Jia Lanpo argues that Chinese scientists Yang Zhongjian and Pei Wenzhong actually found the relics, but Black was given the human fossils to work on while the Chinese were relegated to the other animal fossils.[25] Black dubbed the specimen *Sinanthropus pekinensis*, popularly known as Peking Man. Even though the finds were made in China and not Central Asia, they reinforced Andrews's and Osborn's belief that older specimens would be found in their area. Andrews claimed that Black's discoveries confirmed 'the soundness of the theory of the Central Asiatic origins of primates and man suggested first by Henry Fairfield Osborn in 1900.'[26] While Andrews and Osborn were looking for ancient men in Mongolia, one turned up in Nebraska in the form of an old friend.

Nebraska Man 2

Unexpectedly, Nebraska Man reappeared in 1922. Osborn had been working for years with a geologist and fossil hunter from Nebraska named Harold J. Cook. In February 1922, working in the Snake Creek beds, Cook discovered a tooth that startled him. He quickly sent the tooth to Osborn for examination. Before he did so, Cook compared the tooth against those of other animals as a way of ruling out the possibility of it being a horse, bear or peccary. He was 'confident it must be some sort of primate,' and that it represented an anthropoid-human ancestor.[27] Upon receiving it, Osborn concurred that 'it looks one-hundred per-cent anthropoid.'[28] Interestingly,

Fig. 10 Nebraska Man as the British saw him. This overwrought depiction caused Osborn great consternation and gave Christian fundamentalists an evolution-bashing weapon they still use three-quarters of a century later.

Osborn too ruled it out as being a peccary. He also ruled it out as resembling a living ape. For Osborn it was an example of the 'first Anthropoid ape in America.' The tooth suggested that an archaic anthropoid lived in the Americas and so there was a circumstantial case that archaic people did as well. Osborn called the find *Hesperopithecus* (the ape of the Western world), and told British anatomist Grafton Elliot Smith that it fell between the anthropoid apes and humans.[29] Looking at England's Foxhall Man discovery (a jaw and artifacts found in 1855), Osborn was more convinced that human ancestors would be found in the Americas. He joked to Elliot-Smith that 'I know you will think me half-witted' to say that it was only a matter of time before a fossil human turned up in America that dated to the Pliocene Age.[30]

Later that April Nebraska Man 2 was made public by several newspaper articles and an announcement at a meeting of the National Academy of Science in Washington. Osborn then sent an American Museum team to Snake Creek who produced finds that suggested a forested belt had once stretched from Asia into North America. If it had, it strengthened the early-man hypothesis. Osborn was nervous, however, that the European press was referring to *Hesperopithecus* (he never called it Nebraska Man) as a human ancestor: something he did not believe it to be.[31] His worst fears were realized when, almost immediately after voicing this concern, the ancient dung hit the fan. The June 24, 1922 edition of the *Illustrated London News* ran an article on *Hesperopithecus* with a large double-page spread illustration showing the creature as an archaic, but clearly humanoid, hunter of ancient game, traipsing about the landscape with a crude stick in his hand. Using Grafton Elliot-Smith as an authority, the *London Illustrated News* called the fiì 'an astounding discovery of human remains.' Though Elliot-Smith admitted that nothing was known of the physical structure of the creature, and that the drawing was completely hypothetical, the words were lost beneath the large, exciting picture. Osborn was caught completely off guard by the article, and knew it would cause trouble. He wrote Elliot-Smith saying he could not agree with him that the creature be called an 'ape-man' because there was still so little evidence to do so. Exasperated that the entire affair was beginning to spin out of his control, and that Elliot-Smith was jumping to conclusions, he said, 'I fear that your friendliness and enthusiasm over this discovery has carried you a little further than I would venture to go at the present.'[32] Osborn tried to stem the tide of human ancestry claims with a technical monograph from the American Museum for April and one in *Nature* the following August.[33] Neither helped much.

Osborn argued that *Hesperopithecus* was neither an ape ancestor nor a direct human ancestor, though more related to hominids. He said it was not closely related to Java Man, and was 'therefore a new and independent type of primate.'[34] While all the public pronouncements were being made, William

K. Gregory and other museum staff took a long hard look at the forlorn tooth. Gregory had experience with primate material and compared it with a tooth found at the original Nebraska Man site in 1908, and tentatively concurred with Osborn's findings that the tooth did conform to the types of molars seen in anthropoids and humans. For Gregory, *Hesperopithecus* represented 'an hitherto unknown form of higher primate.'[35]

If Osborn was cautious about making too much of the human connection of the tooth, Gregory was more so. When Osborn later suggested artist Charles R. Knight include a picture of *Hesperopithecus* in a new mural for the hall of mammals, Gregory said it was 'plainly out of the question.'[36] Osborn had gotten over his anger at Grafton Elliot-Smith's reconstruction and wanted to try one of his own. He told Knight to keep working on his studies for the creature and to contact Columbia University zoologist J. Howard McGregor about advising him.[37] McGregor had sculpted several busts of early hominids, but was uncomfortable working on Nebraska Man because of the lack of physical evidence. He annoyed Osborn when he turned down the request to supervise the artist.[38]

By the beginning of 1926 Gregory was having second thoughts about *Hesperopithecus*. Indeed, several minds at the museum were changing.[39] Objections had already arisen by 1923, but the *coup de grâce* came in mid-1925. When Osborn sent an expedition to the Snake River site, they discovered material that was in better shape. Unfortunately, it showed that Nebraska Man was not a man, nor an ape, but an extinct peccary, *Prosthennops*. Ironically, in 1909 when William D. Matthew worked the site with Cook, they too found peccary remains and warned against them being confused with primate remains.[40] Unmasked as a pig, Nebraska Man was quietly dropped from Osborn's repertoire, leaving Gregory to decide what to do about it. The knowledge was originally held quietly, but news 'began to leak out' in 1926 and Gregory found himself being bombarded with questions regarding the fossil. Unable to keep silent any longer, Gregory made an official statement at a meeting of the New York Academy of Science, and in an article for the museum's journal, *Natural History*, which was turned down as being too controversial. Gregory's *mea culpa* regarding the actual nature of the tooth finally appeared in the December 1927 issue of *Science*.

Again Osborn went unfulfilled by human fossils from Nebraska. His evidence was evaporating before his eyes and he was feeling increasing pressure to restate his theories in an attempt to shore up his own beliefs and to convince others of ideas which were being pushed to the side. In a 1922 paper he recapitulated the facts that, for him, supported his position on the mechanics of evolution. In it he tried to convince his audience by sheer force of will. 'No chances or experiments are tried by nature.' He then tried to entice the experimentalists over to his side by holding out rewards for them:

'The experimental botanist, zoologist, biochemist, biophysicist, or geneticist, has the opportunity to win immortal fame by discovering the causes of mechanical evolution.'[41]

As he did with fossil horses, Osborn constructed human family trees in a series of parallel lines which were connected only at the bottom. This technique, called orthogenesis, allowed him to arrange similar animal groups close to one another, but still as separate species. This way he could posit an ancient common ancestor which split to create separate ancestors for both the ape and human lines. He also wanted the human side to be able to then break into more parallel lines, giving rise separately to the Neanderthals, Cro-Magnons and other fossil humans, as well as the modern human races.

Building on the work of William Diller Matthew, Osborn saw humans originating in the Asian highlands. When the heavily forested area began to dry up and the conditions became harsh, Osborn's proto-humans responded with their race plasm to become more human. Historian Peter Bowler points out that the idea of a stimulating environment pushing the proto-men to evolve allowed Osborn to leave the apes behind. The apes stayed in the lush forests while the humans were forced to overcome or perish.[42]

Orthogenesis allowed for predetermined trends to go off in a particular direction unrelated to nearby species, no matter how much alike they looked. Osborn could then safely keep apes, Neanderthals and lesser human types at arm's length where they would not contaminate his race plasm. Bowler argues that Osborn had trouble reconciling all this because, just as he argued the titanothere's trends had led them to extinction, their own trends should push humans to extinction as well. Even the vaunted Nordics were speeding to their doom along their predetermined path, regardless of the environment or their personal efforts.[43] The contradiction put Osborn in a precarious position: he needed orthogenesis to keep the races separate, but that advantage then put limitations on his theory because it suggested struggle and effort were meaningless and that humans were on an inevitable downward slide. He never did clear it up, and instead chose to ignore it.

Matthews's work in *Climate and Evolution* (1914) not only strengthened Osborn's Central Asia hypothesis; it gave him the grounds for saying that the earliest human ancestor, the pre-*sapiens*, had appeared much further back in time, either simultaneously with or even before the appearance of the apes. Osborn dropped the ape connection altogether in 1924 and began to promote his new creation. Though he had been thinking about this for some time – he had begun studying primate evolution as far back as the turn of the century – the changeover seemed abrupt to others. Osborn's friend Grafton Elliot Smith was dismayed by Osborn's change of heart. Elliot Smith assumed all along that Osborn had supported the ape connection. He did see the underlying reason, though, and recognized that Osborn was trying to remove the 'simian stigma' from the human family tree.[44] Osborn needed an early ancestor to get the human line going. He now began referring to the idea, not of an ape-man, but of a Dawn Man.

The idea of creating a hypothetical human ancestor to serve theoretical, political, religious or scientific agendas was not original to Osborn. Others had done, and continued to do, the same. Historian Michael Hammond calls this creature the 'Shadow Man' of paleoanthropology: the supposed ancestor of the human race which was 'lost somehow in the shadows of prehistory' and therefore unable to be proven one way or another.[45] Hammond argues that Marcellin Boule's rejection of Neanderthal as a direct human ancestor, as well as the 'discovery' of Piltdown Man, were examples of this phenomena. As hypothetical constructions, Shadow Men could be used to tweak the Church, as was done in France by socialists eager to oppose religious dogma, or by the conservative Boule who wanted Church support. Osborn used it to keep his theory alive.

The Hall of the Age of Man

Getting nowhere with the scientific community, Osborn took his human family tree to the public. In 1924 he opened the Hall of the Age of Man at the museum. Osborn had always seen the museum as an integral tool in teaching science and culture to the masses. He laid it out in the same way he did his books. Starting with the lowliest forms he worked up higher and higher, building evidence and setting precedents, until he reached the topic all other lessons led to, human evolution.

The hall included several large impressive murals by Charles R. Knight. Osborn wanted the ancient humans to be seen as pristine wildlife that lived naturally and in harmony with their surroundings according to the rules of nature. While they were ostensibly ancient people, Osborn wanted the pictures to be analogous to modern racial distinctions.[46] In the caveman murals the Neanderthals were depicted as robust and tough, their skins dark, their stances brutish, and their attitudes lacking advanced thinking or creativity. Their society had cohesion, but was not as efficiently organized as it could be. The visual moment shows the clan leader fearlessly setting off by himself to run down a wooly rhinoceros that has happened into the clan's domain.

Osborn's directions for the paintings conformed to the most consequential opinion on the status of the Neanderthals at the time. French anatomist Marcellin Boule had done extensive work on the Neanderthals and described their lifestyles as well. In 1908 an almost complete Neanderthal skeleton was found near a site in France called La Chapelle-aux-Saints. Boule's study of the La Chapelle fossils as well as other Neanderthal remains was published as *Fossil Men* (1921). Though Neanderthal had already been characterized as a dim-witted brute, Boule took the model to the extreme. He fashioned an image of

Fig. 11 Charles R. Knight caveman drawing, *c.* 1924. From the collection of the author.

Neanderthals as beast-like creatures, powerful, stoop-shouldered, animalistic and violent in their behavior, communicating by grunts. For Boule the Neanderthals were more ape than man. It is from Boule that we get the classic image of the 'caveman.' Boule, as well as many others, was uncomfortable with the idea of the apparently degenerate Neanderthals as the ancestors of modern humans, especially modern Europeans. Using the orthogenic approach to creating a human family tree, the Neanderthals could be safely pushed off to the side and out of the direct line of descent to modern humans.

By contrast, the Cro-Magnons were shown in their mural as far better organized. They are thoughtful, intelligent and creative as a group. They did not hunt on a whim, but with careful planning and foresight. Their weapons were more advanced, their clothing better made. They had tamed wild animals as allies, and of course were much lighter skinned. They are the picture of Anglo-Saxon, Nordic superiority. Osborn originally wanted the Cro-Magnon cast of characters to be all male. In the end Knight included an attractive, busty young mother in the background suckling her baby and watching the men with an adoring gaze.

The artwork Knight did for this exhibit is some of his finest. The Neanderthal group and an accompanying mural of the Rancho La Brea tar pits are masterpieces that rank with the best of early twentieth-century American art. Osborn liked Knight personally and admired his work to a considerable degree. The museum brimmed over with his pictures for most of the twentieth century (many can still be seen on display there), and Osborn facilitated the sale and promotion of his work to collectors and other museums. It can be argued that Knight owed his career to Osborn's generosity. Nevertheless Knight could be a handful. Osborn had schedules to keep, but Knight worked at a slow pace. He could also be abrasive with museum staff. When his less than speedy progress prompted Osborn to suggest Knight take a female museum artist as an assistant, Knight refused. He said that she was 'so distasteful' to him and that 'she is no artist and never will be.'[47]

Though Osborn claimed the hall was to present empirical evidence of mankind's evolutionary progress, 'and not to confuse the visitor with theories and speculations,' it was arranged to support his human evolution theory. While the affinities between ape and man were shown, it was made clear that they arose from different stocks and were not directly related. The official exhibition guide was written by Osborn, with Gregory stepping in now and then to tactfully edit and explain his boss's less mainstream claims for human ancestry, particularly in the realm of the terrestrial versus arboreal origin question. The years of dogged pestering of anthropologists and anatomists had paid off. The exhibition was packed with original and cast examples of almost all the known human fossils and their artifacts.[48]

Organized in this way, the message was clear: all the races were separate species. Nordic Whites occupied the pinnacle of a line of evolution that began from an early proto-human, non-ape ancestor. In this world everyone knew their place and was happy there. All people had certain abilities and geniuses, but the White race held the most advanced hereditary race plasm, which allowed them to go further and achieve the most.

There were those inside and outside the museum who were uneasy with all the politics involved. Anthropologist Margaret Mead, and even Gregory, squirmed a bit.[49] There were claims that Knight made up all the details, a notion he bristled at. He said he had consulted experts and even actual Neanderthal and Cro-Magnon material in preparing his drawings. Not one to pull his racial punches, Knight shot back at a visitor who said he had made the Neanderthals unrealistically brutish that 'there are many [human] types living on the Earth at present who are far more brutish in appearance … and living habits.'[50] Some objected to what they saw as a certain ungodliness involved. He was criticized for not introducing the concept of a creator into his caveman illustrations, a lack which would harm viewers' morals. Knight replied that God was a personal issue best left to the individual.[51] He was not the only one getting questions over the religious aspects, or lack thereof, of the Hall of the Age of Man. As scholars have noted, belief in evolution in general, and human evolution in particular, was the source of great concern and endless debate throughout the corridors of American thought on both the intellectual and popular level. There was no consensus even within the various camps. In typical American fashion, people accepted or condemned all or part of evolutionary theory as suited their personal tastes. The war over human evolution came down to small pockets of opposing troops who fought brief, but intense, skirmishes.[52] Osborn's creation was about to get caught up in just such a fight.

Evolution and Religion

Christian fundamentalism is often thought of as having emerged in America in the early part of the twentieth century as an anti-intellectual backlash against the rise of Christian modernism and liberalism, the higher biblical criticism, and the teaching of evolution in schools. The work of Ernest Sandeen and George Marsden shows this to be an oversimplification of a very complex issue.[53] Marsden argues that fundamentalism was a development out of revivalism which originated in the early twentieth century, not over political or social concerns, but over what was considered the growth of false Christian doctrine. Sandeen argues that fundamentalism can be traced back to the prophecy movement of the early nineteenth century, and that it broke into the public eye over the modernist and evolution debate of the 1920s.

Christian fundamentalism in America can be generally categorized as a movement roughly held together by belief in the literal meaning of the Bible, opposition to evolution, and a desire to return to traditional forms of worship. Adherents felt themselves pious Christians in a world turning increasingly away from God and proper Christian doctrine. Modernists, on the other hand, held the Bible to be symbolic, not literal, accepted the higher forms of biblical criticism being pioneered in Germany, accepted the teaching of evolution as compatible with Christianity, felt political action could help remedy society's ills, and followed a fluid, liberal view of Christianity. Shailer Mathews, a modernist leader, argued that Christianity had to be 'modernized' by each generation in order to stay viable.[54]

But this generalization can be deceptive.[55] This is in part a result of the grand generalizations made by both sides about the other. Anyone even loosely associated was blindly labeled a fundamentalist or modernist, when in fact the two sides constituted relative minorities within their respective worlds. As a grouping of diverse individuals and denominations – including

Presbyterians, Baptists and others – fundamentalism was not a homogeneous block. There were radical and moderate fundamentalists. There were pre-millennialists and post-millennialists. Northerners as well as Southerners were involved. Some early proponents were not strict biblical literalists, while later adherents did not all subscribe to the same definition of literalism. Not all were anti-intellectuals. In fact, many fundamentalists insisted that they used Baconian induction and Scottish Common Sense philosophy to make scholarly studies of the Bible. This, they said, made their methodology more 'scientific' than the theoretical work of scientists. Their emphasis on gathering and studying facts alone led them in part to their opposition to evolution, which they claimed was theoretically and not empirically based. Not all denominations went over *en masse* to fundamentalism. While many Presbyterians did, the Northern Church had swung more to the liberal side. By 1906 the Northern Presbyterian Church adapted the Westminster Confession to be more Arminian. Salvation became a less rigid and more democratic concept. As historian Sydney Ahlstrom characterized it, 'Are you saved?' came to mean 'Have you decided to be saved?'[56]

The modernists and liberals too – often associated in opposition to the fundamentalists – were not a block. Modernists generally wanted to create a new theology, while liberals wanted a less radical reinterpreting and reforming of the old ways. What is needed here is a sort of unifying theory to link the divergent parts of both modernism and fundamentalism (something beyond the scope of this study) to help clear things up.

The fundamentalist crusade gained notoriety from the attack upon the teaching of evolution in schools. In order to turn the tide, fundamentalist-supported bills were put before the legislatures of twenty states to ban the teaching of evolution throughout the 1920s. Only five enacted them – Oklahoma, Tennessee, Arkansas, Florida and Mississippi. The Bill introduced in Kentucky, spearheaded by the Nebraska politician William Jennings Bryan, strove to outlaw the teaching of human evolution in any school receiving public funds, including universities. The measure was struck down, but die-hard supporters kept on fighting for anti-evolution legislation.[57] Paul Conkin argues that evolution received the wrath of the fundamentalists in part because it symbolized the modernist attempt to create a new God. With the Protestant world in such a state of fragmentation, evolution became the last straw which threatened to bring the Christian God (or Gods, plural, as Conkin suggests) crashing down.[58]

Beginning his career as a Nebraska congressman, William Jennings Bryan (1860–1925) served as an officer in the Spanish American War and ran unsuccessfully for president three times in 1896, 1900 and 1908. The Depression of 1893 led to the virtual destruction of the Democratic Party under Grover Cleveland and allowed for, amongst other things, the election of the Republican William L. Strong as mayor of New York (supported by Madison

Grant and the other Boone and Crocket Club cronies). It had also allowed for Bryan's rise to the upper levels of what was left of the Democratic Party. Seeing an opportunity, Bryan gave his famous 'Cross of Gold' speech in 1896 in which he called for silver to be used as the standard basis of American money. This, he said, would allow poor farmers to pay off their crushing bank loans more easily. This drew support from rural farming areas, and objections from everywhere else. Bryan, like many Westerners, had a deep distrust of the East with its sprawling cities and large foreign populations. He was a staunch temperance man and thinly veiled anti-Catholic, both characteristics which drove away immigrant ethnic voters (the Democratic Party's core constituency). When a Democratic president, Woodrow Wilson, was finally elected again he made Bryan Secretary of State. Beginning in 1914, Bryan helped put much of Central America and the Caribbean into the pockets of American corporations. By the 1920s, he became associated with Christian fundamentalism and stepped to the forefront of the anti-evolution movement.[59]

The same month Bryan's Kentucky anti-evolution bill was killed, he ran an editorial in the *New York Times* excoriating what he saw as the teaching of the heretical belief of Darwinism. Claiming evolution an unproven theory, Bryan said that he opposed its teaching because it was 'harmful, as well as groundless.'[60] The *Times* then asked Henry Fairfield Osborn to write a rebuttal. Osborn said that what Bryan was asking in his article was whether God used evolution as part of his divine plan. Osborn's simple answer was, yes, he did. Science and evolution, Osborn countered, promoted spirituality, not undermined it. He told Bryan the same thing he told his students and other scientists: drop your books and narrow overspecialized thinking and go out into nature and observe it first hand. Doing so would convince any thinking person that evolution did in fact take place and that it was a divinely inspired law which operated slowly and methodically. 'The universe,' he said, 'is by no means a result of accident or chance, but of an omnipresent beauty and order.'[61] Osborn threw not only fossil horses at Bryan, but St Augustine as well, claiming the pagan turned early church father had been an evolutionist. Osborn tried to be civil with Bryan and discuss the matter on a professional level, but Bryan would make that increasingly difficult. The discussion soon turned to name calling and sarcasm. Bryan's name had long since earned a bad reputation in the Osborn household. During the 1896 election Virginia Osborn showed her distaste for the Nebraskan by hoping the country would not add to its woes through 'the disgrace of electing a man like Bryan for president.'[62]

In an effort to show that there was no conflict between science and religion, Osborn signed, along with other scientists and theologians, a statement in the journal *Science* which insisted that science and religion were not irreconcilable, but that they complemented each other.[63] Osborn claimed that the scientific conception of the universe was becoming more spiritual, and that some in the clerical world were coming around to scientific belief. After a French canon named Dorlodot published *Darwinism and Catholic Thought* (1922), Osborn trumpeted that even the Catholic Church had accepted evolution as a fact.[64] The book gave Osborn a whole new positive attitude towards Rome and seemed to prove his contention that the war between science and religion was coming to an end.[65] This made Osborn even more upset with Bryan (upon whom Osborn had focused much of his anti-fundamentalist attention). If the hide-bound Catholic Church could come around, why not the man from Nebraska?

Bryan was not the only fundamentalist who had it in for evolution, or for Osborn. In New York another erstwhile anti-evolutionist set his sights on the activities at the American Museum. John Roach Straton (1875–1929) of the Calvary Baptist Church on West 57th Street was another leader in the anti-evolution movement and took issue with Osborn over the Hall of the Age of Man. Straton was one of a group of outspoken fundamentalist Baptist and Presbyterian preachers who fought a war against the teaching of evolution in the 1920s.[66] He fumed that Osborn and his exhibition were 'poisoning the minds of New York school children by the false and bestial theories of evolution.' Straton wanted the exhibition dismantled or at least to include a reference to the biblical story of creation.[67] Osborn did not want religion introduced into the exhibition because he feared that average people already did not understand the workings of evolution, and religion would only confuse the issue further.[68] Unsatisfied with Osborn's reply, Straton began preaching anti-evolution and anti-museum sermons, even hanging a large banner outside his church lambasting the professor. Straton charged that Osborn and company were committing 'treason against God,' and that taxpayer money should not go to such a den of iniquity as the American Museum. The public tended to side with Osborn. The uproar did not diminish the crowds waiting to see the museum or the hall, and other theologians came to the museum's defense.[69]

The liberal Reverend Harry Emerson Fosdick of the First Presbyterian Church of New York, while not addressing Osborn directly, said that science and Darwinism were not incompatible with Christianity. Fosdick opposed the growing power of the fundamentalists. He was eventually forced out of his position in 1924 by a coalition of fundamentalists which included William Jennings Bryan.[70] Other movement leaders were comfortable with evolution as well. Francis Ellingwood Abbot (1836–1903) was one of the first theologians in America to reconcile religion and Darwinism. John Fiske (1842–1902) blended evolution and idealism.[71] Like Osborn, Fiske believed that Anglo-Saxon Americans were the result of an evolutionary process that could bring prosperity to the world.

While opposing fundamentalism, Osborn also argued that evolution theory was not modernist. Evolutionary thinking, he said, had a long history and was 'in no sense part of Modernism,' but was known by the 'founders of Christianity.'[72] Replying to the counter-editorial, Bryan disparaged Osborn and Princeton biologist Edwin Grant Conklin (who wrote a companion piece to Osborn's), as 'tree men,' equating them to monkeys swinging by their tails and chattering nonsense. He dismissed Osborn's argument completely and claimed Osborn thought the discovery of fossils more important than the birth of Jesus Christ.[73] This must have enraged the devoutly religious Osborn. He was so angry that he had Scribner's publish his 'Evolution and Religion' article as a book the next year in order to refute Bryan's assertions.

While not a biblical literalist, Osborn did not consciously align himself with the modernists. However, as a prominent scientist who was also a devoted Christian who insisted there was no conflict between Christianity and evolution, he assisted the modernists and liberals in their battle with the fundamentalists. An evolutionary theist, he believed that God created the universe and that it ran according to his laws. His insistence on man's free agency being able to transcend predestination fit neatly with the growing Pelagianism of the liberal Presbyterians. Yet he was schooled in, and was a strong proponent of, the same Scottish Common Sense philosophy the fundamentalists claimed as their intellectual methodology.

Where Osborn parted company with the fundamentalists was that he placed his emphasis on nature as a guide to redemption. Also, Osborn was not interested in denominational allegiances or infighting. He called for a basic set of Christian ethics and morals to be adopted, but 'as a man of science,' he said, 'I am not tongue-tied by adherence to any denomination, creed or dogma.'[74] Instead of Church doctrine he looked to the doctrine of nature, where truth was a stable entity. The eternal truths of nature could be read by any layman as easily as by a scientist. Knowledge of evolution, like knowledge of religion, could be useful to people in their everyday lives.[75] For Osborn the language of nature was the same as the language of the Bible.[76]

As the fundamentalists said denial of the truth of the Bible was atheism and that survival of the fittest debased man, Osborn claimed that denial of the truths of nature and science was 'atheism disguised as religion.'[77] He used proofs from the Bible to support his argument (Osborn could quote scripture in the same easy manner as Billy Sunday or Dwight L. Moody) and tweak the noses of the fundamentalists. God tells Job that nature was the result of his power and wisdom. Osborn told the fundamentalists that in this declaration to Job 'are the foundations of true religion. Our moral and spiritual nature is strengthened, not weakened, by the spiritual and moral struggle for existence.'[78] To check the fundamentalists' distaste for Darwin and animalistic struggle further, Osborn countered that 'the moral principle inherent in evolution is that nothing can be gained in this world without an effort.'[79] By bringing science and theology together, he boasted, the youth of America 'will evolve in the right direction' (towards God and salvation).[80]

Osborn, however, was still characteristically murky on some points. When he said that belief in evolution was not 'Modernist' he refers to Augustine making the point that it is acceptable for Christians to believe, but just what he means by 'Modernist' in this case is unclear. Another difficulty in assessing Osborn's position on religion is that he often used religious terms to describe biological function. He used 'constitutionally predetermined' to describe hereditary characters, and referred to the 'race plasm' as being 'immortal.' Though he incorporated a roughly religious model into his evolutionary formula, he rarely talked openly about religion, except broadly. He talked about good Christian behavior and morality, but never discussed predestination, damnation, or even salvation on their own religious terms. W.K. Gregory, one of the people in a position to be able to categorize him, gave up and simply called Osborn an evolutionary theist who kept his deepest feelings about theology private.[81]

Osborn's position can be approached, however, by looking at the question of pessimism and optimism. Taken together, the work of Sandeen and Marsden suggests that eschatology and an emphasis on proper doctrine and right thinking was at the core of the fundamentalist controversy, at least where evolution was concerned. Whereas the fundamentalists' view of humanity's future was pessimistic, that of the evolutionists was optimistic. 'End-time' beliefs viewed humanity on a downward slide to an oblivion from which they could only be saved be Christ. Dispensationalists called for great catastrophes to end each age. The scientific creationist 'day-age' and 'gap' theories relied on catastrophism for geologic change. Osborn was a gradualist. While there were many stripes of evolutionist in the late nineteenth and early twentieth centuries, most held that the human race was moving in a basically progressive and upwardly mobile fashion. This optimistic bent made evolution a false doctrine to the fundamentalists. The work of Paul Boyer (1992) supports this notion. In his study of prophesy belief in America he states that end-time believers 'rejected Charles Darwin's theory of evolution not only because it challenged their reading of Genesis, but because the belief in progress associated with it ran counter to their vision of steadfastly worsening conditions as the end approached.'[82] Osborn's opposition to the fundamentalists can be seen in this context. His view of human destiny was optimistic. He was not a millennarian or a biblical literalist. Anything that undermined his view of human evolution he attacked. In light of the pessimism versus optimism hypothesis, Osborn's work comes into sharper focus. His rejection of fundamentalist attacks, along with his rejection of fossil humans and apes as human ancestors, his creation of the Dawn Man theory, and his efforts in the name of eugenics and immigration legislation, can be seen as an effort to keep downward, degenerate concepts away from the upwardly striving human race.

Undeterred, the Reverend Straton continued his attack, countering with 'evidence' from a fundamentalist geologist from Nevada who claimed to have a fossil shoe print which proved evolution to be a sham. By the mid-1920s Osborn could boast that 'the museum is weathering Dr Straton's attack,' but decided to give in somewhat and make part of the Hall of the Age of Man 'more idealistic.'[83] The verbal sparring between Osborn, Bryan and Straton over evolution was only a warm-up for things to come.

The Monkey Trial

The state of Tennessee passed its anti-evolution law, the Butler Bill, in 1923. The American Civil Liberties Union of New York then enticed Dayton public school physical education and general science teacher John Scopes to act as a test case to overturn the law in 1925. The clash between defense lawyer Clarence Darrow and prosecution counselor William Jennings Bryan over issues of religion and evolution was of immediate interest to Osborn. He saw a chance to get involved when Darrow made the decision to call scientific experts as witnesses for the defense. Scopes was a student of University of Kentucky geologist Arthur Miller, who himself had been a student of Osborn's.[84] In June of 1925 Scopes came to New York to meet with American Civil Liberties Union officials. There, Scopes was introduced to Osborn, Charles Davenport and J.M. Cattell of the American Association for the Advancement of Science. Osborn liked Scopes and gave him advice as well as money to help tide him over.[85] Besides the scientists, Scopes's defense team drew to it a circle of political liberals, socialists, birth-controllers, 'freethinkers, freelovers, and Bolshevists.'[86] It was a situation that annoyed Osborn, and some of the other scientists involved.[87] In the end, for all his support for Scopes and resentment of Bryan, Osborn never did go to Dayton to testify.

The 'monkey trial,' as it was commonly known, seemed tailor made as a public venue for Osborn to pounce on Bryan. He could not have been too busy (as he claimed) for such a historic moment. Several historians have put forward explanations for why he did not testify. Edward Larson suggests that the more prominent scientists bowed out because they did not want to be publicly associated with the radical, socialist, anti-eugenics, agnostic Darrow. Osborn was politically conservative, a eugenicist, and clearly not an agnostic. He opposed socialism, but not fanatically so. He once supported the application of a man to the exclusive Century Club, even though the man was a socialist. Osborn said that if a man were a gentleman, he could overlook his socialist politics.[88] Osborn did confide to Scopes's advisor, G.W. Rappleyea, that he thought Darrow the wrong man to lead the defense.[89]

Wolf and Mellitt suggest another plausible reason: Nebraska Man. By the time of the trial, William K. Gregory already suspected the worst of the tooth. Though as late as August 1925 he was desperately claiming that

Fig. 12 Scopes's defense team, Dudley Malone, Bainbridge Colby, Clarence Darrow, and Henry Fairfield Osborne Neg. No. 311029, courtesy Department of Library Services, American Museum of Natural History.

'Hesperopithecus is not a figment of our imagination,' Osborn too must have had at least some misgivings.[90] It seems logical that Osborn would have loved to beat Bryan and the other Christian fundamentalists over the head with Nebraska Man. (He had already made sarcastic remarks about *Hesperopithecus* being from the same state as Bryan.) Wolf and Mellitt argue that Osborn dropped out of sight because he feared if he introduced *Hesperopithecus* into the proceedings, and it turned out to be something other than a human ancestor, it would backfire and ruin the case for the defense and become a disastrous public setback. In the end, the closest Osborn got to direct involvement in Dayton was quickly to bring out a small book, *The Earth Speaks to Bryan* (1925), which was sent to all the protagonists, and of course put on sale. Fossils, the things he saw as the answers to his questions, continued to undermine Osborn's efforts.

Conclusion

Though he thought Madison Grant's rhetoric a little over the top, by the end of the 1920s Henry Fairfield Osborn's own paranoia had grown to the point where he believed that there was 'a conspiracy … against the Nordic and racial theory' being promoted by those who dismissed it.[91] But it was more than just paranoia. Osborn's theory was unraveling. Beyond immigration and eugenics, other events and ideas had appeared to thwart his beliefs. The 1920s saw an increasing assault upon his ideas. The more threatened his hypothesis became, the more desperate he became to protect it, and himself. To do so he would turn to convoluted explanations and even flirt with the century's greatest evil.

Notes

1. Roy Chapman Andrews, *Ends of the Earth* (New York: G.P. Putnam's Sons, 1929), 4. The story of the Central Asiatic Expeditions has been told in Edwin Colbert, *The Great Dinosaur Hunters and Their Discoveries* (1984), and Douglas Preston, *Dinosaurs in the Attic* (New York: St Martin's Press, 1986) also see the more detailed work on Andrews, Charles Gallenkamp, *Dragon Hunter: Roy Chapman Andrews and the Central Asiatic Expeditions* (New York: Viking, 2001).
2. Andrews, *Ends of the Earth*, 1–197.
3. Ibid., 216.
4. Ibid., 244.
5. The expedition was actually a series of expeditions. The early reconnaissance trips with Andrews and his wife were the first and second Asiatic Expeditions. The trips that the public were aware of, and which gained worldwide attention, were known as the Third Asiatic Expedition, and lasted roughly from 1921 to 1930.
6. *New York Times* (1/2/1921 and 1/16/1921).
7. Andrews, *Ends of the Earth*, 257.
8. See *New York Times* articles from 1921 through 1930.
9. See correspondence between Andrews and Madison Grant, 10/1920, Grant Papers, MAAM.

10. See the Central Asiatic Expedition files, CAAM, for the logistical details and preparations involved in initiating the trip, as well as Gallenkamp, *Dragon Hunter*.
11. Quoted in Douglas Preston, 'A Daring Gamble in the Gobi Desert Took the Jackpot,' *Smithsonian Magazine* (December 1987), 94–105.
12. As Osborn and his Princeton friends did for their trek out west, Andrews and his team armed themselves heavily.
13. Suess, ed., 1899, 'Uberreste Von Rhinoceros sp. Aus der Ostlichen Mongolei,' Mit Anmerkungen Von W. Obrutschen, Ver Handl. KK. Mineral. Gesell. *St Petersburg, Zweite Series*, Band XXXVI, no. 2.
14. Richard Swann Lull, 'The Antiquity of Man,' in *The Evolution of Man* (New Haven: Yale University Press, 1921), 5.
15. *Annual Report of the Board of Regents of the Smithsonian Institution* (Washington: Government Printing Office, July 1890), 463.
16. Ibid., 471.
17. Roy Chapman Andrews, *The New Conquest of Central Asia* (New York: American Museum of Natural History, 1929), 39.
18. Mildred Cable and Francesca French, *The Challenge of Central Asia* (World Dominion Press, 1929).
19. For an overview of other Western explorers in Central Asia see Kenneth Wimmel, *The Alluring Target: in Search of the Secrets of Central Asia* (Washington: Trackless Sands Press, 1996). Also see Edward A. Allworth, 'The Rediscovery of Central Asia: the Region Reflected in Two Collections of the New York Public Library,' *Biblion* (1995), 95–125, and Karl Meyer and Shareen Blair Brysac, *Tournament of Shadows: The Great Game and the Race for Empire in Central Asia* (Washington, DC: Counterpoint, 1999).
20. Alexandra David-Neel, *My Journey to Lhasa* (New York: Harper & Bros, 1927).
21. Nicholas Roerich, *The Heart of Asia* (New York: Roerich Museum Press, 1930).
22. Ibid., 87–134.
23. HFO, 'Journal, China and Mongolia, September–December 1922', box XXV, OSNY.
24. Ibid., 137.
25. Jia Lanpo, *Early Man in China* (Bejing: Foreign Language Press, 1980), 21.
26. Andrews, *New Conquest*, 572.
27. Harold J. Cook to HFO, 3/7/1922, CAAM.
28. HFO to Harold J. Cook, 3/14/1922, CAAM.
29. HFO to Grafton Elliot-Smith, 3/13/1922, CAAM.
30. HFO to Grafton Elliot-Smith, 4/22/1922, CAAM.
31. HFO to Harold J. Cook, 6/5/1922, box IX, OSNY.
32. HFO to Grafton Elliot-Smith, 7/13/1922, CAAM.
33. HFO, 'Hesperopithecus, The First Anthropoid Primate Found in America,' *American Museum Novitates*, 37 (April 25, 1922). See also John Wolf and James S. Mellett, 'The Role of 'Nebraska Man' in the Creation/Evolution Debate,' *Creation/Evolution*, 16 (1985), 31–43, and Stephen Jay Gould, 'An Essay on a Pig Roast,' *Natural History* (June 1989),14–25.
34. HFO, 'Hesperopithecus, the Anthropoid Primate of Western Nebraska,' *Nature*, 110:2750 (August 26, 1922), 281–283; also see *New York Times* (8/17/1922).
35. William K. Gregory and Milo Hellman, 'Notes on the Type of Hesperopithecus Harold Cookii Osborn,' *American Museum Novitates*, 53 (January 6, 1923).

36. William K. Gregory to HFO, 2/3/1924, CAAM.
37. HFO to Charles R. Knight, 3/25/1924, CAAM.
38. Correspondence between HFO and J. Howard McGregor, 6/1925, CAAM.
39. HFO to Harold Cook, 1/27/1926, box X, OSNY.
40. William Diller Matthew and Harold Cook, 'Pliocene Fauna from Western Nebraska,' *Bulletin of the American Museum of Natural History*, 26 (1909), 361–414.
41. HFO, 'Orthogenesis as Observed from Paleontological Evidence Beginning in the Year 1889,' *American Naturalist*, 56 (1922), 134–143.
42. Peter Bowler, *Theories of Human Evolution: A Century of Debate 1844–1944* (Baltimore: Johns Hopkins University Press, 1986), 177.
43. Ibid., 189–195.
44. Grafton Elliot-Smith, *The Search for Man's Ancestors* (London: Watts and Co., 1932), 5–6.
45. Michael Hammond, 'The Shadow Man Paradigm in Paleoanthropology 1911–1945,' in *Bones, Bodies, Behavior: Essays on Biological Anthropology*, ed. George W. Stocking (Madison: University of Wisconsin Press, 1988), 117–137.
46. See correspondence between HFO and Charles R. Knight, CAAM, and the Charles R. Knight Papers, NYPL. Also see Charlotte Potter, 'The Rise to Parnassus: Henry Fairfield Osborn and the Hall of the Age of Man,' *Museum Studies Journal* (Spring 1983), 26–34.
47. Charles R. Knight to HFO, 6/25/1925, NYPL.
48. HFO, *The Hall of the Age of Man*, American Museum of Natural History Guidebook 52 (New York: 1925).
49. 'Agenda for Antiquity' Rainger, 178.
50. Charles R. Knight to *New York Times* [1924?], NYPL.
51. George Becker to Charles R. Knight, 1/15/1922, and Knight's reply, 3/18/1922, NYPL.
52. See other works mentioned in this chapter as well as Jon Roberts, *Darwinism and the Divine: Protestant Intellectuals and Organic Evolution* (Madison, WI: University of Wisconsin Press, 1988).
53. Ernest Sandeen, *The Roots of Fundamentalism: British and American Millenarianism, 1800–1930* (Chicago: University of Chicago Press, 1970) and George Marsden, *Fundamentalism and American Culture: The Shaping of the Twentieeth-Century Evangelicalism 1870–1925* (New York: Oxford University Press, 1980), 25.
54. For the Christian modernist movement see William Hutchinson, *The Modernist Impulse in American Protestantism* (Cambridge: Harvard University Press, 1976) and James Burkhart Gilber, *Redeeming Culture: American Religion in an Age of Science* (Chicago: University of Chicago Press, 1997).
55. Donald Dayton, 'Dispensationalism and the Emergence of Fundamentalism Among American Baptists,' Keynote Address, *Methodism and the Fragmentation of American Protestantism*, Northern Baptist Theological Seminary (September 29, 1995).
56. Sydney Ahlstrom, *A Religious History of the American People* (New Haven: Yale University Press, 1974), 159.
57. See articles in *New York Times* (February 1922).
58. Paul Conkin, *When All the Gods Trembled: Darwinism, Scopes, and American Intellectuals* (New York: Rowman and Littlefield Publishing & Co., 1998), 45.
59. For Bryan's life see Louis W. Koenig, *Bryan: A Political Biography of William*

Jennings Bryan (New York: Putnam, 1971) and Leroy Ashby, *William Jennings Bryan: Champion of Democracy* (Boston: Twayne Publishers, 1987).
60. *New York Times* (2/26/1922).
61. HFO, *New York Times* (3/5/1922).
62. Virginia Osborn to HFO, 8/8/1896, box VI, OSNY.
63. *Science*, l, vii:1483 (6/1/1923).
64. Dorlodot was director of the Geological Institute at Louvain University.
65. *New York Herald* (6/10/1923).
66. Ronald Numbers, *The Creationists* (Berkeley: University of California Press, 1992), 46; also John Roach Straton and Charles Francis Potter, *Evolution versus Creation* (New York: George H. Doran, 1924).
67. *New York Times* (3/9/1924, and 3/10/1924).
68. HFO to George Gray Bernard, 3/26/1924, CAAM.
69. *New York Times* (3/17 and 3/23/1924).
70. Robert M. Miller, *Harry Emerson Fosdick: Preacher, Pastor, Prophet* (New York: Oxford University Press, 1985).
71. Ahlstrom, *Religious history*, 763–770.
72. HFO, *Evolution and Religion* (New York: Charles Scribner's Sons, 1923).
73. William Jennings Bryan, *New York Times* (3/14/1922).
74. HFO, *Evolution and Education: Polemics of the Fundamentalist Controversy of 1922 to 1926* (New York: Charles Scribner's Sons, 1926), 181. Also see HFO, 'Wants Old Religion Taught in Schools,' *New York Times* (1/24/1926).
75. HFO, 'Evolution and Daily Living,' *The Forum*, lxxiii: 2 (February, 1925), 167–177.
76. Ibid., 15.
77. Ibid., 98.
78. Ibid., 99.
79. HFO, *The Earth Speaks to Bryan* (New York: Charles Scribner's Sons, 1925), 25.
80. Ibid., 70.
81. William King Gregory, *Biographical Memoir of Henry Fairfield Osborn* (1938).
82. Paul Boyer, *When Time Shall Be No More* (Cambridge, MA: Harvard University Press, 1992), 96.
83. HFO to J.H. McGregor, 3/20/1924, box IX, OSNY.
84. HFO to Merritt Bond, 1/30/1925, CAAM.
85. There are several histories of the Scopes trial; the most recent is Edward Larson, *Summer For the Gods* (Cambridge: Harvard University Press, 1997) and Conkin, *When All the Gods Trembled.*
86. HFO to Merritt Bond, 1/30/1925, CAAM.
87. Clarence Darrow to HFO, 6/25/1925, CAAM.
88. HFO to Rev. Joseph Duryea, 12/20/1911, CAAM.
89. HFO to G.W. Rappleyea, 8/26/1925, CAAM.
90. HFO to W.K. Gregory, 8/4/1925, CAAM.
91. HFO to Madison Grant, 10/17/1927, box X, OSNY.

CHAPTER SEVEN

Terrible Monkeys

Theodore Roosevelt – soldier, statesmen, adventurer, intellectual, and president – was the embodiment of his friend Henry Fairfield Osborn's ideas about taking one's inbred characteristics and pushing them beyond their limits to achieve more. Osborn held up Roosevelt as the model of American character. In a speech given at the Depression-era dedication of the new Roosevelt wing of the museum in 1931, Osborn told his young listeners that TR's legacy was the lesson to 'live well regardless of personal advantage or disadvantage.' Osborn roused his audience to 'meet … crisis by your own moral and intellectual backbone.'[1] For himself, however, Osborn was not the fearless, intrepid, brawny, outdoor type like Roosevelt or Roy Chapman Andrews, or even a hunting dandy like Madison Grant. While Osborn certainly enjoyed the outdoors, he was not a crasher through the brush, but more a genteel stroller down forest paths. He loved the smell of the trees and flowers, the grandeur of the mountains, and the contemplative uplift it gave him; he just preferred not to get his suit dirty. For all his talk about backbone and effort, he meant intellectual struggle as much as physical daring. In this respect, Osborn was not one of Herbert Spencer's 'henchmen:' those plutocrats and theologians in America who used Spencer's theories to justify capitalist gain, and the weak succumbing to the strong.[2] Osborn's emphasis was on intellectual and spiritual advance, not brute strength. Struggle and effort were cleansing agents that would rid him and his species of an apish taint. He wanted the soap of the past to wash him clean of the dirt of the present, so that he could enter the Eden of the future pure and sparkling.

But dirt, decay and degeneracy were closing in. While battling the roaring twenties, he also had to deal with the rise of Christian fundamentalism, the assault on science and evolution embodied in the Scopes trial, and the embarrassment of the Nebraska Man débâcle as well as trouble within his own family circle. To survive he would have to pull out his last-ditch defensive tool: Dawn Man, the only hero left who could save him. He was slowly being surrounded by chattering fossil cavemen, overspecialized people and disagreeable hordes, a host of terrible monkeys.

As Nebraska Man was dying and the Central Asiatic Expeditions were coming back empty-handed, another hominid was born. Raymond Dart (1893–1988), an Australian anatomist working at the University of Witwatersrand, Johannesburg, South Africa, had been a student of Grafton Elliot Smith. Dart was interested in primate evolution, so in 1924 some of his students brought him a box of primate fossils, including a skull, found at a commercial rock quarry in the nearby village of Taung. After preparing the fossils he was amazed to find that the tiny skull was rather human-like. Thinking he had discovered a new genus of human ancestor, he christened the skull *Australopithecus africanus*, and dated it at over a million years old (far older than anything found in Europe or Asia). Besides having hominid physiology and being African, Taung was everything that Osborn wanted *Hesperopithecus* to be but was not.[3] In a relatively short paper describing the discovery, Dart argued a very Osbornian line. He said that lush jungle lowlands were no place to make humans. The dry uplands had the environment that could fashion apes into men. Only, Dart said, this had happened in Africa not Central Asia.[4] The paleoanthropology community, Osborn included, dismissed Dart's claim for Taung Child. They were too invested in other theories to consider one which shattered them all. The big three of British anthropology, Keith, Smith-Woodward and Elliot-Smith, saw Taung as an interesting, though unimportant, fossil ape unrelated to the human line.[5] Ales Hrdlicka took a brighter view of Taung. He was the first anthropologist to visit the site after the discovery, and though he did not think it a direct human ancestor, he did see it as proof that the early primates had ventured far south. He suggested more excavations be done in order to locate an adult specimen. Alfred Sherwood Romer of the University of Chicago differed: he said Taung was not a chimpanzee or a gorilla. Taking a closer look, by 1930 William K. Gregory thought it did have primitive human-like teeth and was thus not far removed from the apes or humans, whatever it was. In 'A Critique of Professor Osborn's Theory of Human Origins,' Gregory discussed Taung and supported Dart's claim that it was a hominid.[6] Though Osborn and the others did not realize it at the time, or chose to ignore the consequences, the discovery of Taung Child helped to end the paradigm of the Asian origins of humanity. Osborn did not include a cast of Taung in the Hall of the Age of Man, even though he and Dart were acquaintances and Dart would have given him one.[7]

Another African anomaly had appeared in 1921, when a Neanderthal was found in Rhodesia. Since Rhodesia Man was a Neanderthal, and Osborn did not think the Neanderthals ancestors of modern humans, he felt no threat. Dart's claims for Taung were a different matter. For Osborn to say that archaic humans had moved *into* Africa was not a 'paradoxical' idea, but to say human ancestors *came out* of Africa was.[8] That Dart claimed Taung to be a human ancestral ape-man forced Osborn to dismiss the diminutive skull. He had no place in his world-view for an ape-man, and was not the only person to feel that way.

Osborn was particularly worried that the preoccupation with the ape origins of man by some fundamentalists would convince unsuspecting people that evolution meant just that. 'The monkey man theory,' he said, desperately trying to counter that perception in the public eye, 'is pure fiction set up as a scarecrow.'[9] Shortly after the end of the Scopes trial William Jennings Bryan died. (Osborn claimed that he predicted Bryan's demise from all the stress.)[10] His great adversary gone, Osborn thought he could expect less overt opposition, but his human evolution troubles were not going away. He had already sensed trouble back in 1922 and grew frantic over having his theory disintegrate.[11] At seventy-six years of age he was running out of time.

The Scopes trial seemed to mirror the topic of ape–human relations going on in popular culture. Novels and pulp magazines turned out a long line of stories about monstrous ape-men and half-humans. Arthur Conan Doyle had Sherlock Holmes crack the case of *The Creeping Man*, in which a scientist uses monkey gland serum to give himself renewed virility in order to marry a young beauty. *King Kong*, the ultimate ape-as-black-man defiling white-womanhood film, appeared in movie houses just two years before Osborn died. In *King Kong*, the only thing that checks the terrible monkey's rampage is white man's technology (the Empire State Building and armed airplanes). Edgar Rice Burroughs not only created Tarzan, but a host of novels like *The Moon Maid* (1923) and *The Moon Men* (1925), in which ape-like humanoids battled amongst degenerate races, Anglo-Saxon heroes, and beauteous Nordic princesses. Second only to Tarzan as Burroughs's best-known character, John Carter, the fighting man of Mars, found adventure, struggle and interplanetary sex in a series of novels and magazine articles throughout the 1920s, by battling alien degenerates. These works explored the relationship between the savage and the civilized man. Burroughs's cast of characters could fit perfectly into Osborn's or Madison Grant's racial hierarchies. Osborn saw the biological history of man as an adventure not unlike Burroughs's fictional accounts. Both *Men of the Old Stone Age* and *Man Rises to Parnassus* contained much technical material, but were framed as narratives in such a way that they might have been novels 'based on a true story.'

As Osborn tried to widen the gap between man and beast, Sigmund Freud blurred it. Freud sought the savage human mind, the suppressed beast in all of us, controlled by sex and emotion. Freud drew out a creature Osborn claimed did not exist. Despite all Osborn's hard work, the public perception of the relationship between humans and apes was growing. He acknowledged this popular culture trend and saw that it undermined his efforts. 'Let us befriend the Dawn Man,' he said, 'and remove from his reputation the bar sinister of ape descent [which] lingers on stage, in the movies, [and] in certain anti-naturalist literature.'[12]

Dawn Man

After all the setbacks, Osborn would not give up. He could not. He began an all-out campaign of articles, books and radio broadcasts in an effort to prove that he was right about the ancestry of humans. Osborn had hoped

Hesperopithecus would solve his problems, but it only pushed him further from his goal. An actual human fossil had let him down; maybe a hypothetical one would save him. With his pivotal 1926 paper, 'Why Central Asia?', Osborn launched the Dawn Man crusade, and promptly embroiled himself in a public debate over it. Unfazed by the the Taung discovery or the lack of human fossils coming back from the Gobi Desert, Osborn hung on to the Central Asian origins idea and described how he thought the Dawn Men, 'as we [himself, that is] now prefer to call our human ancestors,' developed. The Dawn Men grew up a terrestrial life on the harsh Gobi Plains, forcing them to adapt and become more human and resourceful. The apes, on the other hand, remained in the lush tropical lowlands where the living was easy. That easy life kept them stuck in their arboreal lifestyles. Osborn took the name Dawn Man from Arthur Smith-Woodward's name for Piltdown Man: *Eoanthropus*. This designation, Osborn said, gave the human race 'a line of ancestors ... quite distinct from that of the anthropoid apes.'[13]

Despite the lack of evidence, Osborn still bravely prophesied that the earliest human fossils would be found in Central Asia, though he did hedge his bet somewhat by admitting that 'many decades may ensue before this prophesy is either verified or disproved.'[14] As Misia Landau points out, Dawn Man was the central character in Osborn's epic story of human evolution, a character who had passed through a series of obstacles and adventures like a Greek hero. The Dawn Men began in the trees as distant relatives to the primates. They descended to a life on the ground, stood upright and began to walk in a human fashion. Their brains then grew larger and more creative, allowing them to mature into intellectual beings who went on to become the lords of the earth.[15]

The adventure was laid out in Osborn's most overtly romantic work, *Man Rises to Parnassus* (1927). With it, Osborn was able to bring the circle around. Beginning with Dawn Man, humans had evolved in Central Asia into various separate branches. The most advanced of them were the Aryans. They moved West to become the Nordics and Anglo-Saxons. Achieving an exalted status, they developed a superior culture and society. From there they crossed to the New World to build a mighty nation in a hostile environment through strenuous effort. This effort also sowed the seeds of religion. He speculated that 'the intensified struggle for existence also increased the sense of dependence on supernatural powers and the spirit of mysticism and magic.'[16] It was in America that they reached their peak, where they went from the Old Adam to the New Adam. America was the Eden where they saved not only themselves, but all mankind. There, scientists and intellectuals like himself would become the 'new order of sainthood' to lead the masses to salvation. It was Frederick Jackson Turner's frontier thesis on a planetary scale. As Osborn saw the story of human evolution, it was the Iliad, the Epic of Gilgamesh, La Morte d'Arthur and Horatio Alger all rolled into one grand Bible-like sprawling romance rivaling anything by D.W. Griffith. It was as much wish-fulfillment as science, but it touched on the very essence of humanity. Even his own researches were an adventure. Describing his first entrance into a Cro-Magnon cave site in France, Osborn was as breathless in the retelling as he was in the actual event: 'Again the *élan* on the "cavernist" carried us through all difficulties into the final chamber.' He continued, 'on the ceiling … we perceived in faint relief the painted figure of the sorcerer.'[17] But the adventure of man was tinged with pathos and a dire warning for the future. 'Have our youth gained in reverence and faith,' he asked rhetorically, 'our women in modesty and love of family? Are we following in the very footprints of our psychologists who have lost the soul, the mind, and finally consciousness itself?'[18]

Osborn had come out publicly and stated that the story of human development had no room in it for an ape-like ancestor. More than his stance on eugenics and immigration, this issue brought him attention. A public battle erupted, pitting him against virtually the entire paleontological and anthropological establishment. His primary protagonist in the drama was his own right-hand man, William K. Gregory. The two had grown close since Gregory became the boss's assistant. Osborn respected Gregory's work and though he tried to get Gregory to do research that supported his interests, Osborn gave him free rein to follow his own path. Consequently, Gregory's view of human evolution was very different from Osborn's. His extensive work on primate physiology had guided him to a view that accepted the close relationship between humans and other primates. This difference of opinion never caused any personal animosity between the two. Though they were from different social backgrounds, Osborn often referred to Gregory as 'fidus.' He also promoted Gregory's career and always complimented him in public whenever possible.

For years Gregory would swallow hard and grin nervously when Osborn pronounced his latest theories. He might disagree with Osborn's assertions, but he chose to do so in private. That reserve was shattered in 1926 when Osborn began arguing Dawn Man in earnest. Gregory already suspected the professor had decided to remove the apes in 1920 when Osborn asked him to review a draft of a paper he wrote on the subject.[19] Gregory stuck with the monkeys. In 1924 he did a study of the newly discovered fossil primate from India, *Dryopithecus*, and concluded that it was 'man's and gorilla's lineal forefather.'[20] This creature, with its arboreal history, was the antithesis of the terrestrial Dawn Man.[21] *Dryopithecus* brought man closer to the apes as Osborn tried to push them away.

In public, Osborn said the reason for his abandonment of ape lineage was a question of anatomy and psychology. Apes, he said, do not move at all like humans. Their hands and feet, while superficially similar, were not so on a deeper level. He also argued for differing psychologies. Again using the horse as the model, he said if we look at various species of horse we see that 'the ass has a desert psychology, the horse has a forest psychology, and the zebra has a psychology of the plains.'[22] They have a mindset, a certain predisposition for their environment and lifestyle. This concept of 'psychology' crosses over into primates as well. He said that ape psychology and human psychology differ to an astonishing degree and thus the two groups could not be biologically related.

Osborn argued that wherever you found horse fossils you would find human fossils, because the two followed the same migratory behavior, something the early primates did not do. Making clear the importance of will-power and personal drive, Osborn said that the pro-humans' special 'psychology,' race plasm and wanderlust made them 'the architects of their own destiny as the horses and titanotheres' were the architects of theirs.[23] Gregory called the fear of ape ancestry 'pithecophobia,' and continued to rebut Osborn point for point.[24] Unwilling to hear such things, Osborn told Roy Chapman Andrews to hold out in China because he would find fossil human ancestors in the Gobi Desert and that they 'will not be apes, but true pro-men.'[25] Gregory accused Osborn of completely misreading the evidence, and asked in exasperation how anything could be learned, as Osborn insisted it could, by comparing horses and primates.[26]

Osborn also pointed to the foot, hand and thumb of both primates and humans as a contrast to show their differences. He charged that an arboreal lifestyle – living in the trees and swinging from branch to branch – could not have led to a human upright walking structure. Only a terrestrial upbringing in a harsh environment where the organisms had to struggle to survive could have done that. Gregory countered that an arboreal lifestyle was just the sort of beginning humans would need.[27] Swinging from branch to branch called for grasping hands which in turn led to higher brain power and bipedal locomotion.[28] He disagreed with the old linear orthogenic model of human evolution favored by Osborn and others, and found no validity to the idea of predestined factors. He favored Darwin's chance and variation.[29] Gregory also felt that the Scopes' trial played a significant role in Osborn's sudden flurry of Dawn Man activity.[30]

Osborn fervently wanted Gregory to believe as he did. Gregory refused, but hated the public feuding with his beloved mentor and friend. Fidus told Loulu that he did not want to oppose her husband, but his conscience would not let him sit back and do nothing as Henry Fairfield pronounced the ape–human origin connection a 'myth.' To Osborn himself he pleaded, 'I think you know that I do not go contrary to you from choice and it is with profound regret that I feel morally bound to take issue with your views.'[31] He was devoted to Osborn and the argument caused him much personal distress. He also feared the divide would be exploited by Christian fundamentalists.[32]

While the American press kept an eye on the proceedings, the British media watched as well. They asked why Osborn had abandoned the apes in

Fig. 13 Osborn is thanked by the apes for leaving them alone. *Kansas City Times*, January 10, 1930.

Neg. No. 271655, courtesy Department of Library Services, American Museum of Natural History.

what seemed to them a sudden manner. They had assumed him to be a Darwinian.[33] Osborn felt the British press was misrepresenting him, particularly after his retirement speech as president of the American Association for the Advancement of Science. In the speech he talked at length about Dawn Man. He thought the British accused him of attacking Darwin.[34]

Throughout 1926 and 1927 the two engaged in this intellectual game of tennis in articles, scientific meetings and newspapers. In 1930 Osborn even went on the popular Collier's Radio Hour to promote Dawn Man. In a folksy, FDR-like address, Osborn poked fun at himself by saying how much he enjoyed a recent political cartoon where a grateful ape thanks him for being released from human kinship. He finished the talk by assuring his audience that it had become his life work to convince people of his position.[35] The *New York Times* ran a full-page illustrated article about human origins which cited both men and their respective positions.[36] While most of the scientific establishment in the United States and England sided with Gregory, Osborn did have his defenders. The commissioner of Labor Statistics in Washington thanked Osborn for his opposition to ape ancestry because he felt it supported immigration restriction and other racial attitudes.[37]

The Christian Right too must have felt quizzical gratitude for this 'evilutionist' who, as they did, said man did not come from monkeys. One self-righteous reader asked him, 'So you are the religio [sic] scientist who was worsted in the debate?'[38] On the other hand the editor of the *New York Forum* told Osborn proudly that his Dawn Man remarks had gone over well in Europe, particularly in Germany. The Germans, he said, 'are excited about every word you utter.'[39] Excerpts of his Tertiary Man address turned up in newspapers in New York, Boston, Pittsburgh and San Francisco. The eugenicist biologist David Starr Jordan told Osborn that reading the Dawn Man material reinforced his belief that Osborn was right about a lot of things, while Harvard anthropologist Earnest Hooton told Osborn he would be happy to work with him on a detailed study of Dawn Man, but in the end he was 'on the side of Dr Gregory and the great apes.'[40]

Grafton Elliot-Smith, while dismayed by Osborn's change of heart, saw the Dawn Man debate as positive. With Osborn saying such things in public, it forced others like Gregory to come out and explain why the professor was wrong. This would help to explain evolution to those who did not understand it.[41] Though Osborn was always cheerful and optimistic in public, the stress took its toll. In the summer of 1928 he suffered a nervous breakdown which took him several weeks to recover from.[42]

Osborn and the Occult

Osborn's brother, William Church, watched the debate from the sidelines and then wrote to congratulate him. Willie told Farfie that he never personally gave the ape-origin question much consideration, feeling it was not that important an issue. He cared little 'whether the monkey was my grandfather or my second cousin.'[43] He was just so proud of his older brother that he wanted to let him know. There were many, however, who cared deeply about this question, and not all of them were scientists or theologians.

As the expeditions to the Gobi Desert had aroused interest in the occult community, the Dawn Man debate also attracted scrutiny by devotees of Madame Blavatsky and theosophy. The Russian mystic's thoughts on human evolution, first expounded in 1888 in *The Secret Doctrine*, were surprisingly in line with Osborn's. He had some knowledge of Blavatskian theology since on a train in Nebraska, he had once whiled away the hours engaged in a lively debate on theosophy and spiritualism with some fellow passengers. He later wrote that he found the experience 'rather entertaining.'[44] There were those who saw in Osborn's Dawn Man pronouncements scientific validation for something they had been saying all along: man was not descended from the apes, but arose independently through mystical and unknowable divine agencies. Occult adept G.L. Metteru suggested Osborn read Blavatsky's work to see for himself how compatible the theories were.[45] Another, Basil Crump, a Briton living in China, concurred. He pointed out that Osborn seemed to be saying, as Blavatsky did, that not only were the apes not the root of humanity, but that the apes were an offshoot of the human line.[46] Crump, a student of Eastern philosophy, had written a book on Eastern views of evolution and the development of the brain and felt Osborn was coming close to an oriental understanding of human origins.[47] Using an interpretation of Osborn's own pronouncements, Crump attempted to show how close the Western paleontologist was to an Eastern view. It was noted that Osborn's estimate for the age of man 'agrees with the archaic Eastern teaching' on evolution.[48]

While he certainly held some notions that could be called mystical, Henry Fairfield Osborn was not a mystic. That his human evolution theory was similar in some respects to Madame Blavatsky's was a consequence of their mutual objective: an ape-free ancestry and proof of spiritual salvation.[49] There is no evidence to suggest he had any serious interest in the occult, though he did have a few intriguing brushes with it.[50]

In January 1920 Osborn attended an unusual gathering at the home of the canon of Manhattan's church of St John the Divine. He was there with a group of New York Episcopalian ministers to hear a talk by physicist and 'psychical researcher' Sir Oliver Lodge (1851–1940). Lodge did research into psychic phenomena, including spiritualism, and believed that an individual's spirit and memory survived after death. He talked about his ability to contact the spirits of the dead and claimed this was not only possible, but could be proved scientifically.[51] The canon, G.W. Douglas, asked Osborn to come along as an impartial scientific observer who had no interest in the occult. The attendees were asked to keep the meeting a secret, but word leaked out.[52] It was reported that 'profound interest in the subject … was evident throughout the meeting.'[53] Osborn apparently did not see anything that impressed him or made him believe that communication with the dead was possible.

Following the Dawn Man articles a medium from New York, Herman Behr, contacted Osborn with an unusual way to determine which side was right. Behr asked Charles Darwin to clear up the confusion. Despite the fact that the father of natural selection was dead, Behr claimed to be in contact with him. If Osborn wanted to ask Darwin any questions, Behr could be an intermediary, and get the 'answer in a reasonable time.' He could also contact a host of other notable personages from Savonarola and George Washington to Franz Liszt.[54] Osborn's reaction to this unique opportunity can only be guessed at, since he never replied.

Osborn's mother thought that the spirits of the dead could watch over the living, but he seems to have had little interest in the idea. The empirical part of him may have felt the evidence lacking; the romantic in him may have preferred to leave the question 'unknowable.' Either way, by the 1930s he knew his own end was drawing near. He once told William K. Gregory to start preparing his own work for posterity so that he could leave something of value behind. Osborn was practicing what he preached. He worked tirelessly on an enormous study of the origins of the elephants and began wrapping up loose ends in his research to formalize, once and for all, his theory of evolution. His science was no more accepted now than it had been forty years before. In fact, he was even further from universal support than he had ever been.

On The Trail of Ancient Man

Despite having made such a point about the Central Asiatic Expeditions' search for fossil humans, Roy Chapman Andrews did not bring an archaeologist with him until 1925. That man, Nels Nelson, was a staff archaeologist with the museum. He had worked closely with Osborn on *Men of the Old Stone Age*, supplying technical advice and ideas. Nelson, however, was dubious about the prospects of finding early humans in Central Asia. He questioned some of the stone implements, or eoliths, found there as well as those found in Europe. He had no problem with the idea of eoliths, but thought that the criteria for accepting them as such were not rigorous enough. He also expressed his doubts about these supposed human artifacts to one of Britain's leading anthropologists, Arthur Smith-Woodward. Unfortunately, Smith-Woodward was not impressed by Nelson's argument.[55] Undisturbed by Nelson's doubts, Osborn still wanted him to go. As optimistic as ever, Osborn told the archaeologist that 'Paleolithic man in Central Asia is eagerly awaiting your arrival,' and that his 'mission is to hunt our ancestors in their original lair, to prove that their sojourn in France was after a long preliminary period of primary education on the vast roof of the world.'[56] The hunt would have spectacular results, just not those originally envisioned by the team.

Walter Granger's initial discovery of fossils at Iren Dabasu was an auspicious beginning. The team went on to find, not armfulls of fossils, but truck-loads. The Gobi Desert turned out to be a vast storehouse of fossil remains rivaling that of the American West. Andrews and his men found mammals and dinosaurs alike. They found small ones and huge ones, quadrupeds and bipeds. Besides many unknown species of dinosaur – from the diminutive *Protoceratops* to the feisty *Velociraptor* – they found the first-known dinosaur eggs.[57] They found horses, rhinos and titanotheres (including the largest land mammal ever known, *Balucatherium*, which was as tall as a two-story house) and other mammals. Central Asia certainly seemed to have been a center of radiation for an entire world of organisms. In this respect the Third Asiatic Expedition was a huge success. Andrews filled the museum's coffers with an unprecedented collection of fossil animals. The press ate it all up and Andrews, Osborn, and the American Museum became household names around the world. There was, however, one conspicuous absentee from the list of fossils being sent back to New York: man.

Though human fossils were not turning up, the expedition did discover signs of past human habitation. Artifacts and burials were found which put the crew on edge, suspecting they had finally found what they were looking for. Unfortunately, they found Neolithic Man, not Pleistocene Man as they had hoped. Labeled the 'dune dwellers' by Andrews, they were some of the earliest people to live in Central Asia several thousand years ago. While the dune dweller's history could reveal much about Central Asia's early civilization, they were modern anatomically and as such could shed no light on human origins. It was a major disappointment.

When no human fossils were forthcoming after the 1925 season, Nelson wrote Osborn his regrets. While he felt that they would not find evidence of early humans in the Gobi, he held out hope that it might be found in China.[58] Walter Granger had wanted to look for human fossils up the Yangtze River between 1921 and 1923, but heavy bandit traffic in the area kept that from happening. Andrews wanted to send Nelson to China because the Jesuit paleontologist and philosopher Pierre Teilhard de Chardin told him that fossil humans could be found there.[59] Teilhard De Chardin, who like Davidson Black spent some time with the Andrews expedition, was uncovering Neanderthal remains in the Ondos Desert of China, but the China Geological Survey had asked the American Museum team not to prospect in China as they saw it as their domain. An agreement was reached: Andrews stayed in the Gobi Desert and never did prospect for human remains in China. That would be left to others. For his part Nelson argued that Central Asia 'never offered any outstanding material attraction to primitive man' due to its inhospitable climate. Though Central Asia certainly had an early culture (the dune dwellers), he saw no evidence for the region having been inhabited by archaic people or proto-humans.[60]

As early as 1926 the writing was on the wall for the Central Asiatic Expeditions. China was in chaos. Looting and banditry were so bad it was dangerous to go out on the streets. The United States government was suggesting its citizens pack up and leave. The American China Gunboat Fleet was more active than ever, rescuing people and defending itself against attack. On an outing to shoot moving film of the landscape, Andrews and some of his men found themselves caught between warring factions and a hail of gunfire. Making it back to the city unharmed, they joked about the poor marksmanship of the Chinese soldiers, but their car was riddled with over one hundred bullet holes. They knew they were lucky to be alive. Andrews admitted to Osborn that 'we really had no business to be where we were … you can be perfectly sure it won't happen again.'[61]

Despite his optimism, Andrews knew the end was near. He asked, 'What [do I do] in case we have to give up work in Mongolia?' Like the last kid who refuses to leave the ball field even though it is raining, Andrews lamented to Osborn in tragic frustration, 'I am not ready to settle down to a curatorial job yet. I have too much restless energy. I am too full of health and strength and enthusiasm.'[62] He was not giving up on his boss's theory either. He resisted looking for human fossils in Burma or Java. Osborn told Andrews 'You are now right in the country where our ancestors struggled for their existence and my daily prayer is that you have the good fortune to hit upon the remains … the jaw or the skull or the femur will tell the whole story.'[63]

It was growing darker in China. While Andrews disliked the Communists, he also had little love for Chang Kai Shek's Nationalists, whom Andrews called 'the devil's chosen swine!' He raged at the chaos that was costing him precious fossil-hunting time, as well as the lives of countless Chinese men, women and children. With seething anger he spit that the ex-warlord Chang Tso-Lun may have been a bloodthirsty pirate, 'but by Jove, he knew how to keep the Chinese in order which can only be done by cutting off heads!' To make matters worse, the Chinese government was confiscating his fossils and household furniture as relics of the Chinese people. They also claimed that Andrews was spying, exploring for oil and selling opium. He was eventually given the material back after he paid bribes to officials. Exasperated, he told Osborn that 'it is a pretty difficult thing to deal with a people who can only talk parrot-like about their "sovereign right" and "unequal treaties" [while] half of them don't know what the words mean.'[64]

The breaking point was reached in 1930 and the expedition was decommissioned and brought home. An unexpected result of the expedition was that it not only changed the view of Mongolia, but Mongolia itself. Andrews had proven that the Gobi Desert could be successfully traversed with the use of motor transport. He hoped that the country's pristine condition would be preserved, but within a few years the desert reverberated to the sounds of hundreds of cars and trucks trading goods and exploring for mineral resources. As author Douglas Preston points out, the ancient landscape and culture of Mongolia was irrevocably changed by Andrews himself – something he did not want to do.[65]

Though he found no fossil humans, Andrews still believed Osborn's 'prophesy' would be proven correct. They had learned much about the natural history of Central Asia and its role in mammal evolution and radiation, but the real prize still seemed within reach. To appease Osborn, Nels Nelson lamented that 'the Central Asian Expedition was well on its way to this imagined prehistoric paradise and the archaeologist's chief regret is that we failed to find it.'[66] Osborn was not about to give up. Events were closing in around him, however, that were out of his control. The failure of the Gobi missions to find extinct humans was being compounded by the rise of the Christian fundamentalist movement, events in Africa, Tennessee, Nebraska, and right in his own neighborhood of Manhattan. When the team returned for the final time to New York there was a sense of disappointment despite the great success. They had made magnificent discoveries, but the great prophesy regarding humans had not come true. The oracle had spoken unwisely: Adam had not rediscovered paradise. To make matters worse, the other most important thing in his life, his family, was suffering as well.

Things Fall Apart

Where Henry Fairfield Osborn had pedaled with his son Perry through what he saw as a jewel-like city, he now saw darker things lurking in the corners. His city was gone, his theories rendered irrelevant, and his family was disintegrating. In 1928 Osborn could say that 'all … family news is good.' Josephine, her brother Fairfield Jr and their families were heading to Europe. Perry, who had been battling depression, returned from spending time with Virginia and her family at her home in Arizona feeling better, and Osborn himself was looking forward with Loulu to watching the construction of the new Theodore Roosevelt Memorial entrance on the museum from their Fifth Avenue apartment. He was also putting together plans for a new book that would 'rout for all time the "warm, tropical jungle theory" as the birthplace of man.'[67]

The rosy picture did not last long. The Great Depression put a squeeze on the entire Osborn clan. The core family fortune was weakened, but held, even though Henry Fairfield had to stretch William Henry's money to support his adult children.[68] By the end of 1929 Perry's marriage broke up. His wife Anne could take no more, went to Reno, Nevada, got a divorce and almost immediately remarried, leaving Perry with custody of their two children.[69] Perry confided 'the whole story' of the break-up to his mother. Part of the problem stemmed from Perry having injured his back playing polo. Having damaged several disks in his spine, he was constantly in pain, which made it difficult for him to concentrate. The failure of divorce did not go down well with a father who believed in both the superiority of his blood line and struggling to achieve greatness.[70] Osborn told his son he should get remarried as soon as possible so that he could still provide a stable home and stepmother for his children, but Perry's life was a shambles. He had seemed to show the least promise of all the Osborn children. Fairfield Jr was also showing signs of strain. His father had warned him that 'your intellect is not disciplined by hard times, it is following the path of least resistance.'[71] In 1922 Josephine, who thought her father the greatest man on earth, worried that her siblings could never live up to their father's accomplishments. She sighed, 'It's too bad none of his children take after him in brilliancy.'[72]

As bad as Perry's troubles were, they were nothing compared to the heartbreak that was about to befall the family. In their forty-ninth year of marriage Loulu's health took a downturn. She had been suffering poor health for a decade. The illness, combined with a growing depression (about what is unclear), rendered her physically and emotionally weak. He tried to cheer her by taking her to the opera in April. That worked temporarily, but she soon relapsed with bronchitis.

After fifty years at the museum, Osborn put in his last official work day there on May 9, 1930, after being given indefinite leave to work on the elephant volumes. When he went home, Loulu was worse and by August the end was in sight. She was so bad that Virginia and Josephine rushed home to take care of her. Henry spent every day at her bedside at Castle Rock, reading to her and comforting her. They were 'sad, sad, days,' as Henry Fairfield could do nothing but watch his wife slowly die. Finally, on August 26, 1930, as the clock ticked and Virginia and Josephine wept uncontrollably, he sat on the edge of the bed (the same bed her father had been born in) and 'held my own darling's hand–while her soul passed.'[73]

She was the light of his life. She stood by him through thick and thin, her support never wavering. Theirs was a fairytale romance in which they sent each other love letters every week (sometimes every day) of their long marriage. Her death devastated him and he never recovered from it. He told Roy Chapman Andrews that Loulu's death was 'the greatest crisis of my life.'[74] (Andrews could sympathize somewhat: he and his wife Yvette had just divorced.) Osborn could suppress his emotions over the deaths of almost anyone, but not his beloved Loulu. Normally a prolific letter and diary writer, he was only able to send a letter to Andrews in China. After that his personal diary is blank for several days.

Compounding his anguish, Osborn made a point of remembering the anniversary of the death of his son Gurdon Saltonstall (he remembered all the death dates). Though it had happened over thirty years before, he was still haunted by it. He blurred that death with the untimely demise of Virginia's boy, his grandson Osborn. As it had always been, the only thing that saved him from a complete breakdown was work. The struggle of work fortified him and helped him keep going. He continued to put in long hours on the elephant project, and it helped, but he could not constantly 'drown my loneliness and sorrow in hard work' and the anguish often grew unbearable.[75]

With the Great Depression still wreaking havoc on the boys, their father asked them to come back and live at Castle Rock until better times returned.

Osborn was desperate to shore up the clan. He took on the financial responsibility of paying the educational costs of at least ten of his grandchildren.[76] These setbacks were not supposed to befall a superior bloodstream such as the Osborns. He had done extensive research into his personal racial ancestry (though as with much of his research, others, particularly Madison Grant, contributed). He believed that his name was an anglicized version of Asbiorn, and that it was Scandinavian for 'divine bear.' The Asbiorns, he believed, were Norman warriors and the 'divine ancestors of the Norse race.'[77] His Nordic American ancestry was important to him, as was his desire to be considered a 'native.' He complained to Edward Poulton that the term 'foreign' should be dropped from his membership in the British Royal Society. 'I find it impossible to feel,' he said, 'that I am a foreigner in any sense,' even in England.[78]

Posterity mattered to him. He wrote in his diary, 'secure the family!'[79] Consolidating the tribe could make the estate 'absolutely secure.' Frantic to hold things together, he asked himself if it would ever 'be possible to restore the spirit' of his wife to Castle Rock. Unfortunately, 'Perry seemed absolutely blind' to their situation. It was no wonder; the youngest Osborn boy was depressed and nothing seemed to be bringing him out of the condition.[80]

Teetering on the edge of mental and physical breakdown himself, in February 1932 Henry Fairfield Osborn left on a round-the-world cruise on board the heavy luxury yacht *Stella Polaris*. The trip, besides relaxing his nerves, opened his eyes towards the cultures he encountered, particularly in the South Seas and Indo-China. The ship made numerous stops in Polynesia, Malaysia and Australia. He marveled at the cultures and people that seemed in no way degenerate or inferior. He visited the pyramids at Angkor, and archeological sites in Singapore and Saigon. The trip, and the anthropological data and ideas it generated, restored his spirits. Ever the teacher, during the cruise he took to giving the other passengers impromptu talks on the various places they visited. He told Josephine buoyantly that, 'I have made the voyage' by entertaining everyone with his knowledge and wit.[81] Upon his return the family held a large reunion and Thanksgiving dinner at Castle Rock which included seventeen second cousins, eight first cousins and three grandparents, and a host of others.

Osborn was still deeply concerned about his son and worried that Perry's future was all but gone. Henry Fairfield wanted Perry to take over control of the family the way his own father had wanted him to take over the railroad decades before, but Perry just wasn't interested. The situation was ruining Osborn's already fragile health. The recuperative effects of the *Stella Polaris* trip vanished and he began to suffer constant indigestion and other ailments. He spent the days at Castle Rock writing, walking to Loulu's grave and decorating it with her favorite flowers from the garden, and struggling with the crushing emotional pain and loneliness. By December 1932 he gave up on the idea of Perry saving the family and decided to do it himself. Perry then left his New York law practice and moved to Miami. The other lawyer in Osborn's life, however, was still hard at work.

Conquering a Continent

After the success of *The Passing of the Great Race* Madison Grant spent the better part of the next two decades working on immigration legislation and generally promoting eugenics. Grant first announced his intention to write another book in July 1930.[82] Osborn was happy Grant was writing again, but warned him against 'depreciating other races' while trumpeting the Nordics. In fact, Osborn suggested Grant leave other races alone altogether to 'set forth their own greatness.'[83]

Ignoring Osborn's warning, Grant's *Conquest of a Continent* (1933) is one long rant against non-Nordics. Where Nordics had built the country, he argued, non-Nordics were ruining it. He lashed out at them all: Blacks, Jews, Poles, Russians and Italians. He covered much of the same territory as in *The Passing of the Great Race*, but with greater emphasis on the elimination of the unfit.

Grant believed that steps needed to be taken to save the nation. He suggested registration for all people living in the country so that deportation would be easier. Naturalization should be suspended for at least a year. Sterilization must be instituted and carried out wholesale on undesirables. The unfit – feebleminded, alcoholics, drug addicts, the handicapped and Bolsheviks – must be curbed in their ability to leave 'behind [the] posterity of their debased type,' and a 'council of perfection' would need to be put together to see the measures carried out smoothly.[84]

Grant believed that due to the altruism and 'fatal sentimentalism' of Whites, the 'lower orders' were asserting themselves in Europe and around the world. Grant saw the self-determination of colonial people as a threat to social order. Imperialism had a positive effect upon native peoples in the same way slavery had a positive effect on Blacks in America. Throughout Africa, for example, tiny minorities of Whites were ruling vast populations of Blacks and keeping them contentedly in their place. The natives, he said, were better off this way since without a guiding white hand, they would remain savages. 'Hindus,' who had no such guide, had 'ruined' Polynesia. He argued democracy 'simply will not work on Asiatics.' With these things in mind, Grant warned, Whites of the West, especially Americans, must be prepared to meet the onslaught. They would have to set up laws and statutes to keep out undesirables, remove those already here, keep Blacks in their place and 'bear our share of the white man's burden.'[85]

As he did with his first book, Grant received support from varied quarters. Ullrich Meier, a German national and officer of the Universal Book Mart (an international book dealer with offices in Montreal, Toronto and New York), wrote to Scribner's to tell them that sales of Grant's book would do well in the fatherland. In a matter-of-fact way he stated that 'The support, moral and ideal, of the [German] government may be taken for granted, as, acc[ording] to my knowledge, Madison Grant is supporting the views and contentions of the National Socialist Government. I feel sure that … Dr Goebbels' Ministry of Enlightenment should be very pleased to point out these works.'[86]

Grant's work had always been embraced by eugenicists and Nordic supremacy advocates in Germany, including members of the Nazi hierarchy.[87] Allen Chase argues that Adolph Hitler based part of *Mein Kampf* on *The Passing of the Great Race*.[88] When Leon Whitney (1894–1973), a eugenicist who worked on the congressional immigration committee with Grant, tried to impress Grant with a letter from the Führer, Grant gleefully produced his own in which Hitler called *The Passing of the Great Race* 'his Bible.'[89] Grant's first book was considered essential reading for German race theorists, and was partially reprinted in the German newspaper *Berliner Börsenzeitung* in February of 1938.

Having become by the 1930s a eugenics writer with a growing reputation and many connections, Grant took an unexpected path by turning his back on his oldest, most influential colleague. In 1916, Henry Fairfield Osborn had written the introduction and rewrites to Grant's first book, and was the influence that pushed Charles Scribner to publish it. In 1932, he was asked to write the introduction for *The Conquest of a Continent*. Osborn felt the early draft was full of wonderful material, but thought the writing was lacking. He told Grant that if the draft was just a set of rough notes they were 'priceless.' If Grant was putting it forward as a finished text, 'it requires much more work.'[90] Osborn suggested dropping the terms inferior and superior. The *Stella Polaris* voyage had reinforced in Osborn that all 'races' had their own traits of greatness. If only Grant could see what he saw, Grant would 'abolish the terms inferior and superior.' He insisted Grant should eliminate such prejudice from the book to make its message more forceful. In the end, even with these drawbacks, Osborn thought the new book 'great.' He confided to his diary that while he liked the ideas of Grant's book, he felt they were 'often poorly expressed.' Despite his qualms, Osborn dutifully put together a flattering introduction for his old friend.[91]

Editor Maxwell Perkins, however, was not happy with the results. Though disappointed in the text, Perkins said that it was at least 'better than the one he did for *The Great Race*.' Accordingly, Perkins wanted to cut Osborn's introduction altogether. It would not be that big a loss, he felt, because of Grant's now increased reputation.[92] Grant never questioned Perkins's opinion, and agreed that Osborn's introduction was not worthy of being included. In the end it was, but as a compromise to keep from offending supporters, not as an act of respect for the man without whose help Grant would have been just another rich lawyer ranting over race questions to his rich club friends.[93] Unaware of the backstabbing, Osborn continued to be a friendly supporter of Grant until the end.

Aristogenesis

Osborn still had his own theory to defend. While working on the elephant project he hit upon an idea. If you could tell precisely at what point in the past a fossil elephant had lived – based upon the layers of tooth enamel – you could extrapolate the age of the human fossils found near it.[94] He claimed that this enamel evidence showed the modern human line to be at least one million years old. The press picked up his pronouncement and saw an inherent comedic value; several satiric cartoons of cavemen and elephants appeared. One pundit, Harry Acton, chortled nonsensically that according to Osborn, 'the family dentist is our uncle.'[95]

Still determined to prove, or at least convince others of, his human evolution theory, Osborn repackaged it one last time. He still insisted that morphology and the study of structure could reveal deep insights about evolution. The elephants, he argued, had advanced beyond the stage of development achieved by the titanotheres, particularly in their grinding teeth. Just as he felt the titanotheres were integral to understanding mammalian evolution, he was now convinced that the elephants were even more so. He was also trying to play catch-up with the geneticists. He said that science was unable 'to comprehend the hereditary germ plasm' without adding the study of morphology to experimental biochemistry and biophysics. He was not through with the old versus new biology, and tried to save paleontology's place in it.[96] While by now he accepted natural selection as an important factor in evolution, he was still convinced that it was not the central mechanism. He rebuked the experimentalists that for all their work they had experienced 'utter failure to explain the origin of species.'[97] He said the work of traditional paleontologists, like himself, had undermined the notion of 'created evolution,' but supported what he called 'creative evolution.' That organisms adapt to their environment creatively was a process he now called aristogenesis. This process operated on a series of 'rectigradations' which allowed the organism to adapt purposefully.

Rectigradations, according to Osborn, are the predestined factors in the race plasm which give the phyla its special powers and abilities. Aristogenesis, he said, was the 'creative origin of the best and fittest.'[98] Creativity was like will-power: it could be called upon to overcome obstacles. To him, one of the things bringing on society's downfall was a loss of creativity. He saw children having their creativity wrung out of them in school, where they were forced to learn by rote instead of experience. They were taught to parrot answers instead of think for themselves, as he had been taught. His study of the past had revealed to him that a lack of creativity doomed species and cultures to extinction. The cavemen had no schools or forced conformity. They learned naturally by living in nature where the tough environment forced them to develop their creativity or perish. That was how they achieved their advance and that was why he admired men like Theodore Roosevelt: men who would not be reined in, who used their creativity, struggled and were rewarded. What the country needed was more men like him.

Osborn's creative evolution idea was similar to that of French philosopher Henri Bergson (1859–1941).[99] Like Osborn, Bergson was unsatisfied by both Lamarck and Darwin. In his *Creative Evolution* (1907) Bergson argued that evolution was pushed by an *élan vital* (a sort of will-power) that compelled an organism to overcome adversity. Like Osborn, he refuted mechanistic processes and argued for a life of the body and one of the spirit. While not agreeing with Bergson completely, Osborn admired much of his work and incorporated some of his philosophy about creativity into his own. The term 'creation,' Osborn insisted, was not meant in a theological way, but in the sense of an organism being able to create its own future in original, or creative, ways. Organisms could combine their predestined rectigradations with a fluid will-power to make a unique response to the environment. Aristogenesis was a refined form of what he had been arguing all along.

In his last important paper, Osborn tried one more time.[100] In explaining aristogenesis he said there was no element of chance in the creative process. Borrowing from the work of Albert Einstein and Neils Bohr, he said that 'life is a "quantum" of adaptive action, reaction, and interaction.' In a last effort to make his theory understood, he recapitulated all his work and how it led to his final conclusion. He rolled out all his 'discoveries' and theories, all of which he claimed were built upon direct observation of the fossil record. Lined up, this work led to the latest breakthrough in evolutionary thinking. Quoting himself in the third person, he referred to an 1890 paper where he claimed there was an evolutionary factor which had been hypothesized, but unknown, which had to exist as a sort of unifying factor of evolution. He had finally discovered that unknown factor and it just happened to be aristogenesis.

Osborn said the aristogenes were the characters that appeared (using a more modern designation) in what he now called the 'geneplasm' and which slowly 'evolved into an important functional service.'[101] So here it was: aristogenesis was the core of the race plasm. The only piece of the puzzle he did not have – and no one would have – was how the aristogenes appeared in the first place. He insisted that this was unknowable, in the way God was unknowable. In a speech to St Bartholomew's Church, Osborn insisted that his study of aristogenesis suggested to him that the modern human line and its ancestors, as well as the human soul, were immensely old. They were so old they could not possibly be ape-related.[102]

Regardless of his efforts, Osborn's ideas were not being embraced, not even by his friends. In 1931 Edward Poulton organized a large-scale celebration of Charles Darwin, as well as the British Association for the Advancement of Science, called 'One Hundred Years of Evolution.' Speakers were scheduled to discuss various aspects of the history of evolutionary biology, including the latest breakthroughs and ideas. Poulton asked Osborn to be part of the festivities, but to his chagrin Osborn was asked to speak only on the fact of evolution, 'in this case not touching on causes or principles, but just the evidence that evolution is a reality.'[103] Poulton did not want Osborn to introduce any of his quirky, God-driven, non-Darwinian theories into a celebration of Darwinism.

Osborn dutifully swallowed his pride and did as Poulton asked. Josephine and William K. Gregory went along on the trip and turned it into a pleasant tour of England. Osborn placated himself by saying that it was probably better for him not to discuss aristogenesis or any of his other ideas because the British press would only accuse him of bashing Darwin, the way they had before.[104]

Nothing would be easy for Osborn. Not only was he denied a chance to speak about his theories before a gathering of leading evolutionary biologists; he was being ignored by history before he was even part of it. In 1935 Poulton published a paper on the history of evolution including a survey of everything known about it. He did not mention Henry Fairfield Osborn or any of his work. Peeved by the slight, he told Poulton sarcastically 'not to think for a moment that I have the slightest personal feelings about your omission of even the barest allusion to my thirty-five years of direct observational work … on the process of evolution.'[105] It was a bitter pill to swallow that for all his work in the field of evolutionary studies, he did not rate even a footnote.

While little notice or support for aristogenesis came from the scientific community (by now he expected little), Osborn continued to receive glowing words from some quarters. Eugenicist Harry Laughlin praised aristogenesis and began to incorporate it into his own work on races and race horses.[106] George Ashley, the state geologist of Pennsylvania, wrote excitedly to him that reading Osborn's aristogenesis work convinced him that 'the age of man is about over and that time is ripe for an age of superman.'[107] Besides a few well-wishers, the final academic acknowledgment for his work came from a source all too familiar with supermen.

Osborn and the Fascists

The rise to power of the Nazis in 1933 did not fill Osborn with dread the way it did others. He saw the Nazis as a positive force in Germany. In his limited understanding of the situation he thought the Nazis were helping to raise Germany from the chaos of the Great Depression and the crushing effects of the World War to put it back on the road to greatness. He saw, not a criminal regime with evil intent, but a beaten Nordic people using their race plasm and scientific advancements to overcome great obstacles.

Osborn's connection with fascist Europe came during the *Stella Polaris* voyage. He was impressed by how the Black Shirts had reorganized Italy and put it back in shape. He told Gregory that while in Italy he hoped to meet with Benito Mussolini to discuss the situation.[108] Osborn thought the way Il Duce had transformed the country was 'marvelous.'[109] Unlike that of Madison Grant, Osborn's work was not embraced by the Nazi hierarchy. He was, however, recognized by the German academy. In August of 1934 the University of Frankfurt gave Osborn an honorary doctorate in natural history.[110] (He was the first American given such an honor during the Nazi era.) After going to Germany to receive the award in person, Osborn toured the Fatherland. Impressed by what he saw, he said prophetically that observing the youth of Germany was 'to realize that the world will have much to deal with in the not too distant future' because of them.[111] The following November the journal *Natur Und Volk* celebrated Osborn's work on evolution, the Nordic race and eugenics. The article ended with a long quote from Osborn himself in which he returned the praise for German society. He said that Germany was misunderstood by the world and that he would personally do his best 'to correct unfavorable impressions that now prevail.'[112]

In addition to such well-known Nazi sympathizers as Henry Ford and Charles Lindbergh, other Americans also had positive things to say about German government. Osborn's philandering, socialite friend Poultney Bigelow (by then eighty years old) visited Germany the next year and even spent time with the old Kaiser. Bigelow thought the United States could learn much from Germany's new order and seriousness. He felt America had become too concerned with good times and behaved like a 'national nightclub' of partying and lax morality (this from a man made famous in New York society for his extra-marital affairs and high living). In addition, the former United States ambassador to Germany, Alanson Houghton, thought Hitler was 'working the thing out' and improving German status and society.[113]

For those who saw the Nazis in positive terms, their own prejudices allowed them to be unconcerned with the growing racist, anti-Semitic, tendencies of the Nazis towards non-Aryans. Osborn may have been bamboozled by the Nazis' early embrace of science as a centerpiece of their ideology and the many nature-oriented youth movements blossoming throughout Germany.[114] On his trip, made in his late seventies, Osborn saw the smiling face of Nazism, the happy youth frolicking in the forest, learning from nature and getting in touch with their Nordic heritage. He did not, or chose not to, see the darker aspect. Nazi statements about youth education, rebuilding the country along moral lines, and an insistence that the future lay in applying biological principles to society was something Osborn had been advocating for America all along. He told Harry Laughlin that 'my recent trip to Germany, Austria, and Italy and my continuous observation in the United States convince me that of all the modern social movements, eugenics is far the most important.'[115]

Gregory could not believe Osborn's remarks. He was hurt and disturbed by them, and in response resigned from the Galton Society. He could no longer belong to the organization because of the 'tendency of a number of our members to take sides with the Hitler government.'[116] What made things worse, in his eyes, was that it was all being done under the guise of patriotism and science.

Osborn tried to mollify Gregory by insisting that the truth about Germany was not getting out because the 'Jewish press' in America was distorting the picture. He told Harvard anthropologist Earnest Hooton the same thing. To Osborn the *New York Times* was 'strongly anti-Nordic.'[117] Referring to Ambassador Houghton's pro-Nazi position, Osborn said that 'his views exactly coincide with my own.'[118] Gregory's reaction gave Osborn pause to regret his words, if only because he feared losing his friend and successor at the museum. Osborn wanted Gregory back on his side and back in the Society. He also feared that he had personally wounded Gregory and that their friendship had been put in jeopardy because of it. 'If I have erred in anything I have said about Germany,' he begged Gregory, 'I hope you will forgive me.'[119] Gregory could not. What upset him was that he sensed Osborn had been taken in by the Nazis' lip service to science and building a society around a eugenic ideal. Gregory knew that Osborn had allowed his judgment to be swayed by the idea that a nation was following the sort of path he himself had advocated for national betterment. Gregory could never forgive Osborn for his remarks, even though very few of them had been made in public. Osborn never did not get a chance to clear things up with Gregory or see the real face of Nazism. The rift only added to the pain of a heart already shattered beyond repair by the death of his wife.

The End

In April 1934 Osborn officially retired from the museum. Sad to see 'The Professor' go, Roy Chapman Andrews told Osborn that he owed everything to him, and that the museum would never be the same without his presence. In a touching statement about Osborn's omnipresent countenance in his office, Andrews said that 'when the time comes … when you no longer sit at the head of the table, I feel I shall never want to go into that room again.'[120] With his retirement, Osborn retreated to the quiet of his library at Castle

Rock. Though ill, he kept up a busy schedule of work. He was determined to struggle on to get his evolution theory clearly written down for posterity. Like some silent movie gunslinger he told Poulton that he suspected he was 'destined to die with my boots on.'[121]

His doctors were not so happy about his cowboy work schedule. They insisted he slow down and rest. His heart and circulation were all but gone and his blood pressure and pulse were abnormally high, so the doctors put him on a mild sedative. They insisted he stay in bed or he would drop dead. He would have none of it and continued his work.[122]

From his library at Castle Rock Osborn kept working and corresponding as vigorously as he could. He discussed science and politics. He continued to rail against president Franklin Roosevelt's new income tax programs, which he felt unfair to the rich. He tried to get an article published in the *Women's Home Companion* on 'What a Girl Can Do.' Taken from the title of a series of articles by women's rights champion Carrie Chapman Catt, the answer was anything she wanted (though he still insisted the greatest thing a woman could do was create healthy children). The editor turned it down because the topic had already been covered by others, and the article was far too long and rambling for a magazine.[123] He also now spoke out publicly about birth selection over birth control. Echoing his stance on immigration, he called for 'not more, but better and finer representatives of every race.'[124] Promoting birth selection, and still advocating eugenic measures for the good of society, Osborn contributed to *The Literary Digest*. Excerpting his speech from the Third Eugenics Congress – like the second, held at the museum – he said that birth control was 'death-dealing rather than … birth encouraging.' Accompanying the article was a photo of Osborn sitting on the porch of his summer home looking frail and tired holding a copy of the titanotheres volume. At his feet was Josephine, nattily dressed in the latest fashion and staring adoringly up at her father.[125]

By November 1935 he seemed to rally and get some strength back. On the night of the fifth he spent a pleasant evening with members of Perry's family, having dinner and relaxing in front of the fire. Later he went to bed feeling tired, but happy. As was his way all his life, he rose early the next morning and went to work in the library. His desk was covered with the elephant manuscript as well as the latest plans for the Roosevelt memorial. Sometime before eleven, as he sat at his desk he took one last look out the window at the autumn colors of the Hudson Valley landscape. He then leaned forward and quietly died of a heart attack. He was seventy-eight years old.[126]

Notes

1. HFO, 'Cornerstone Ceremony, Theodore Roosevelt Memorial' (10/27/1931), mss in box XX, OSNY.
2. James Moore, 'Herbert Spencer's Henchmen: The Evolution of Protestant Liberals in late Nineteenth-century America,' in *Darwinism and Divinity*, ed. John Durant (New York: Basil Blackwell, 1985).
3. A good description of the Taung discovery is in Roger Lewin, *Bones of Contention: Controversies in the Search for Human Origins* (New York: Touchstone, 1987), 47–84.
4. Raymond Dart, 'Australopithecus africanus: The Man-Ape of South Africa,' *Nature*, 2884:115 (February 7, 1925), 195–199.
5. Arthur Keith, Grafton Elliot-Smith and Arthur Smith-Woodward, 'The Fossil Anthropoid Ape of Taung,' *Nature*, 2885:115 (February 14, 1925), 234–236.
6. Noel Boaz, 'American Research on Australopithecine and Early Homo, 1925–1980,' in Frank Spencer, ed., *A History of American Physical Anthropology* (New York: Academic Press, 1982), 240–241.
7. That Osborn turned down Dart's Taung cast is interesting when compared with the great lengths he went to to get other fossil casts. It is total speculation on my part, but it makes one wonder if Osborn saw the same implications for Taung as Dart did. It is possible that he suspected Taung was just what Dart said it was and so kept it out of the museum exhibit as an out-of-sight, out-of-mind approach to damning evidence.
8. HFO to Grafton Elliot-Smith, 12/6/1921, CAAM.
9. *New York Times* (7/12/1925).
10. HFO to Edward Poulton, 8/26/1925, box X, OSNY.
11. HFO to W.K. Gregory, 7/12/1922, CAAM.
12. HFO, *Man Rises to Parnassus* (New York: Scribner's Sons, 1927), 75.
13. HFO, 'Why Central Asia?', *Natural History*, 26:3 (1926), 269.
14. Ibid., 269.
15. Misia Landau, *Narratives of Human Evolution* (New Haven: Yale University Press, 1991).
16. HFO, *Man Rises to Parnassus*, 81.
17. Ibid., 95.
18. Ibid., 192.
19. W.K. Gregory to HFO, 11/30/1920, box 72, Vertebrate Paleontology Archive, American Museum. Hereafter VPAM.
20. *New York Times* (4/15/1924).
21. *New York Times* (1/9/1927).
22. HFO, 'Report Before the Galton Society,' 2/2/1927, folder 6, box 72, VPAM.
23. HFO, 'Recent Discoveries Relating to the Origin and Antiquity of Man,' *Science*, lxv:1690 (May 20, 1927), 481–488.
24. W.K. Gregory, 'Two Views in the Origin of Man,' *Science*, lxv:1695 (June 24, 1927), 601–605.
25. HFO to Roy Chapman Andrews, 10/6/1927, CAAM.
26. W.K. Gregory, 'Is the Pro-Dawn Man a Myth?,' *Human Biology*, 1 (May 1929), 153–165.
27. William K. Gregory, 'A Critique of Professor Osborn's Theory of Human Origins,' *American Journal of Physical Anthropology*,' 14 (1930), 155.
28. William K. Gregory, 'Studies on the Evolution of Primates: Phylogeny of Recent and Extinct Anthropoids with Special Reference to the Origin of

Man,' *Bulletin of the American Museum of Natural History*, 35 (1916), 258–355.

29. Ronald Rainger, 'What's the Use?: William King Gregory and the Functional Morphology of Fossil Vertebrates,' *Journal of the History of Biology*, 22:1 (Spring 1989), 103–139.
30. W.K. Gregory to D.M.S. Watson, 1/3/1927, box 18, Gregory Papers, CAAM.
31. W.K. Gregory to HFO, 4/21/1927, Gregory papers, CAAM.
32. W.K. Gregory to Lucretia Osborn, 12/17/1927, box X, OSNY.
33. *London Morning Post* (12/12/1929).
34. HFO to Sir Richard Gregory, 1/13/1930, box 71, VPAM.
35. HFO, 'The Discovery of Dawn Man,' Radio Address, 2/2/1930 (New York: American Museum of Natural History, 1930).
36. *New York Times* (5/11/1930).
37. Ethelbert Stewart to HFO, 12/28/1929, box 71, VPAM.
38. Tom Follett to HFO, 12/18/1929, box 72, VPAM.
39. Henry Goddard to HFO, 1/29/1930, box 72, VPAM.
40. David Starr Jordan to HFO, 5/4/1929, and Earnest Hooton to HFO, 5/29/1929, box 72, VPAM.
41. Grafton Elliot-Smith, *The Search for Man's Ancestors* (London: Watts and Co., 1931).
42. HFO to Madison Grant, 8/21/1928, CAAM.
43. William Church Osborn to HFO, 12/30/1929, box 71, VPAM.
44. HFO to Lucretia Osborn, 7/21/n.d., box XI, OSNY.
45. G.L. Metteru to HFO [1/1930?], box 71, VPAM.
46. *North China Standard* (3/13/1930).
47. Basil Crump to HFO, 6/6/1930, box 71, VPAM.
48. Typescript, possibly by Basil Crump, 1930, box 72, VPAM.
49. A major difference in their thinking was that Blavatsky saw time and space in a more classical or pagan cycle of death and rebirth to a higher never-ending plain while Osborn was more Augustinian in his belief that existence traveled along a straight line with a finite end.
50. In 1934, someone at the Scribner's company inexplicably sent Madison Grant a copy of Madame Blavatsky's *Isis Unveiled*. Grant returned it saying he wasn't interested, and had not asked for it. Madison Grant to Scribner's Company, 6/20/1934, box 67, Author files I, Scribner's Archive, PU.
51. Lodge wrote a series of books including *Evolution and Creation* (1926), *Science and Human Progress* (1927) and *Why I Believe in Personal Immortality* (1928).
52. Howard Robbins to HFO, 1/23/1920, and G.W. Douglas to HFO, 1/23/1920, box IX, OSNY.
53. *The Churchman* (January 31, 1920).
54. Herman Behr to HFO. 5/22/1931, CAAM.
55. Nels Nelson to HFO, 11/12/1924, CAAM. Smith-Woodward used eoliths from England and the Continent to support his human evolution ideas.
56. HFO to Nels Nelson, 2/18/1925, box X, OSNY.
57. Osborn named both *Velociraptor* and *Tyrannosaurus*.
58. Nels Nelson to HFO, 11/3/1925, CAAM.
59. Roy Chapman Andrews to HFO, 7/6/1924, CAAM.
60. Charles Berkey and Nels Nelson, 'Geology and Prehistoric Archaeology of the Gobi Desert,' *Novitates* (3/18/1926).
61. Roy Chapman Andrews to HFO, 4/30/1926, CAAM.

62. Roy Chapman Andrews to HFO, 7/7/1927, CAAM.
63. HFO to Roy Chapman Andrews, 6/27/1928, CAAM.
64. Roy Chapman Andrews to HFO, 8/31/1928, CAAM.
65. Preston, 94–105.
66. Andrews, *New Conquest*, 601.
67. HFO to Roy Chapman Andrews, 6/27/1928, CAAM.
68. HFO to Edward Poulton, 5/9/1933, box X, OSNY.
69. HFO to Edward Poulton, 3/14/1930, box X, OSNY.
70. HFO diary, 1/12/1930, box XXVI, OSNY.
71. HFO to Fairfield Osborn Jr, n.d., box XI, OSNY.
72. Josephine Osborn to Lucretia Osborn, 1/11/1922, box IX, OSNY.
73. HFO diary, August 26, 1930, box XXVI, OSNY. Also see her obituary in the *New York Times* (8/27/1930).
74. HFO to Roy Chapman Andrews, 9/15/1930, CAAM.
75. HFO to Edward Poulton, 3/11/1931, box X, OSNY.
76. HFO to Ned Frost, 7/15/1933, box X, OSNY.
77. For Osborn genealogy see box I, OSNY. The quote is from Madison Grant to HFO, 1/16/1933, same location.
78. HFO to Edward Poulton, 5/16/1932, box X, OSNY.
79. HFO diary, 1/1/1932, box XXVI, OSNY.
80. HFO diary, 8/25/1932, box XXVI, OSNY.
81. See February to June correspondence, box X, OSNY. For the quote see HFO to Josephine Osborn-Coogan, 5/6/1932, box X, OSNY.
82. Madison Grant to HFO, 7/3/1930, CAAM. Osborn advised Grant to tone down 'your lack of preference for other races!': HFO to Madison Grant, 7/14/1930, CAAM.
83. HFO to Madison Grant, 7/12/1931, CAAM.
84. Madison Grant, *Conquest of a Continent* (New York: Scribner's Sons, 1933), 351.
85. Ibid., 353, 356.
86. Ullrich Meier to Scribner's, 7/23/1935, box 67, Author Files I, Scribner Archive PU.
87. It was Grant himself who pushed to have foreign-language translations prepared. A French edition of *The Passing of the Great Race* was published in 1926. A German edition was brought out by the notorious German racial hygiene publisher Julius Friedrich Lehmann in 1925. *The Conquest of a Continent* translation was issued by German publisher A. Metzner in 1937.
88. Chase, *Legacy of Malthus*, 169.
89. Leon F. Whitney, unpublished MSS autobiography, 1973: 205, Whitney Papers, APS.
90. HFO to Madison Grant, 1/14/1931, CAAM.
91. Ibid.
92. Perkins to Grant, 8/22/1933, box 67, Author Files I, Scribner Archive, PU.
93. Grant to Perkins, 8/19/1933, box 67, Author Files I, Scribner Archive, PU.
94. HFO to Edward Poulton, 4/8/1931, box X, OSNY. Also HFO, 'Biological Inductions From the Evolution of the Probocidea,' *Science*, 76:1979 (December 2, 1932), 501–504.
95. Harry Acton, 'On the Gangplank,' *New York American* (10/9/1931).
96. HFO, 'Paleontology Versus Devriesianism and Genetics in the Factors of the Evolution Problem, ' *Science*, 73:1899 (May 22, 1931), 547–549.

97. HFO, 'The Principles of Evolution Revealed by Paleontology,' *Nature*, 3239:128 (November 28, 1931), 894–897.
98. HFO, 'The Discovery of Creation in Evolution,' Address before St Bartholomew's Church, New York [1931?], box XVI, OSNY.
99. Bergson's sister was married to the occultist S.L. McGregor Mathers, founder, along with Aleister Crowley, of the influential occult society, The Golden Dawn. See Elic Howe, *Magicians of the Golden Dawn* (Wellingborough: Aquarian Press, 1985), 64.
100. HFO, 'Aristogenesis, the Creative Principle in the Origin of Species,' *American Naturalist* (May–June, 1934), 193–235.
101. Ibid.
102. HFO, 'The Discovery of Creation in Evolution,' Address before St Bartholomew's Church, New York [1931?], box XVI, OSNY.
103. Edward Poulton to HFO, 6/10/1931, box X, OSNY.
104. It is ironic that Osborn was reduced to such a role in the celebration. That Darwin's home, Downe House, was turned into a shrine was largely due to Osborn's efforts. For years he had rallied support and raised funds to buy back the house and restore it. He also donated a clutch of original Darwin papers to the collection. See letters in box X, OSNY.
105. HFO to Edward Poulton, 10/28/1935, box X, OSNY.
106. See the Laughlin–Osborn correspondence for 1934 in the Laughlin Papers collection, Pickler Memorial Library, Truman State University.
107. George Ashley to HFO, 3/1/1933, CAAM.
108. HFO to William K. Gregory, 4/5/1932, Gregory Papers, box 13, CAAM.
109. HFO to Madison Grant, 8/24/1934, CAAM.
110. *New York Times* (8/24/1934).
111. *New York Times* (9/14/1934).
112. *Natur Und Volk* (November 1934), 435–439.
113. *New York Times* (5/10/1935).
114. For Nazi science see Robert Proctor, *Racial Hygiene: Medicine Under the Nazis* (Cambridge: Harvard University Press, 1988), 27–39 as well as Ute Diechmann, *Biologists Under Hitler* (Cambridge: Harvard University Press, 1996).
115. HFO to Harry Laughlin, 10/3/1934, Courtesy of the Harry H. Laughlin Collection, file D-2–4:11, Pickler Memorial Library, Truman State University.
116. William K. Gregory to C.G. Campbell, 5/6/1935, Raymond Pearl Papers, Galton Society folder, APS.
117. HFO to Earnest Hooton, 1/22/1934, CAAM.
118. HFO to William K. Gregory, 5/25/1935, Gregory Papers, box 13, CAAM.
119. HFO to William K. Gregory, 3/23/1935, Gregory Papers, box 13, CAAM.
120. Roy Chapman Andrews to HFO, 4/13/1934, box XXI, OSNY.
121. HFO to Edward Poulton, 1/14/1935, box X, OSNY.
122. Max Wolf to Charles Clark, 10/4/1935, box X, OSNY.
123. Willa Roberts to HFO, 7/30/1935, box XVI, OSNY.
124. *New York Times* (5/4/1932).
125. HFO, 'Eugenics As the Cure for All the Race's Ills,' *The Literary Digest* (September 10, 1932), 22–24.
126. *New York Herald*, obituary (11/7/1935).

Who Saved Osborn's Soul?

Osborn's death was the end of an era at the museum. The grand, world-wide attention-getting expeditions of the past were over. A marble bust of him was put up in the Hall of Mammals, and life moved on. While eulogized as a great leader, Osborn's Dawn Man and Central Asia theories were filed away and forgotten. Though he always claimed not be a Darwinian, Osborn was remembered in the press as 'Darwin's successor.' His funeral was held at Garrisons and he was buried in the family plot in the cemetery of St Philip's Church next to his parents and his beloved Loulu. Memorial services were held in New York City.[1]

Despite their public feud over Dawn Man, Osborn wanted W.K. Gregory to succeed him at the helm of the museum, but the board of trustees gave the job first to Trubee Davison, then to Roy Chapman Andrews who had been cooling his heels in a curatorial and public relations position since his return from Asia. Andrews's tenure in the front office was short and rocky. A great field man, Andrews made a mediocre administrator. By the time of World War II he had been asked to step down. With his second wife, Wilhelmina 'Billie' Christmas, he spent the rest of his life reliving the Gobi expeditions and telling the story to anyone who would listen or he could corner. He published several more books on the adventure, but they were just retellings of the same tale. (Some books lifted entire passages word for word from previous volumes. One was a novelized version called *Quest in the Desert* (1950) with himself as the hero.) He spent his last days dreaming of Iren Dabasu and Ulan Bator, and died quietly in California in 1960. After being passed over for president, William K. Gregory continued to work at the museum as curator emeritus of fishes and comparative anatomy. He also continued to teach at Columbia, where the year before he retired in 1945 he was elevated to Osborn's old chair as DeCosta Professor of Zoology. He lived out the rest of his life in upstate New York, and died in 1970.

Madison Grant outlived Osborn by only two years. He had suffered for years from crippling arthritis, yet continued to rant and rave until the end. He died in 1937, still a bachelor and still angry at the world. With the loss of his great patron, Charles R. Knight's career went into eclipse. He illustrated books, several of which he wrote, and his work still hung in museums around the country, but he received few new commissions. His paintings were reproduced innumerable times, and became the basis of many film dinosaurs and monsters from *The Lost World* to *King Kong* to *One Million* BC, usually unaccredited.[2] He spent his final years in a Manhattan apartment fulminating anathemas against moral decay. He hammered angrily into his typewriter about how he hated modern art. In 1935 he told the *American Magazine of Art* that the work they were highlighting – particularly that of Henri Matisse – was 'loathsome.'[3] He was also unhappy with the Metropolitan Museum of Art over their exhibition of Dada and Surrealist art. He claimed to be one of the few 'respectable citizens who loathe and abhor these monstrous and degenerate explosions of artistic nothingness.'[4] His life and work slipped into obscurity until he was rediscovered by a new generation in the 1980s, helped along by his granddaughter, Rhoda Khalt, who had inherited and preserved a large collection of his original work.

Somewhere between dreams and reality lie people's actions. In order to support his evolutionary theory, Osborn entered the discourse on race and ethnicity. To back up his contention that organisms had certain innate abilities and characteristics, he separated them, not only into distinct races, but into separate species. This led to these distinctions being viewed as those of superiority and inferiority. As had been the case, anthropology in general, and then human origins in particular, was used both directly and indirectly to give a stamp of approval to bigotry and intolerance. The Victorian belief that science was neutral and based upon fact alone added weight to the notion that racial distinctions were 'natural' and that treating one group as biologically and culturally superior to another was 'scientifically' sound. Osborn would have argued that he was looking out for the betterment of all mankind. He did not think himself a bigot, a racist, or anti-Semite. So caught up in his study of human origins and evolution, the pursuit of his own salvation, and the idea that he was intellectually honest, he was blinded to the harm he inadvertently helped promote.

Humans have always treated each other in ways based upon their origins. Where a person is from was believed, and in many ways is still believed, to say a great deal about that individual's potential as a human being. What family, neighborhood, city, country, or ethnic background a person traced their ancestry to has often influenced how they are treated by the wider world. Osborn thought that being able to trace humans back to their first appearance as humans in Central Asia could give order and stability to the world. Understanding how humans originated and evolved would impact upon questions outside of the merely biological, reaching to the social and cultural. Osborn's lasting contribution to the modern dialogue on race is that he helped to establish the role of human origins in the discussion. Where different human groups originated could be used as a guide, he believed, to how those groups should be treated and where they fell in the social hierarchy. Little did he realize that knowledge of humanity's African origins would be used in a similar fashion, but with different goals in mind. Many modern human-origin theorists, as well as sociologists and civil rights supporters, argue that knowledge of our shared African origins can be a unifying factor in world civilization: claiming that as we all sprang from the same well, we are all deserving of the same equal, ethical and just treatment, and that despite our superficial differences we are part of the same family.

While the basic notion of human origins continues as a framework for discussions of larger ideas, Osborn's Central Asia hypothesis and his Dawn Man argument never caught on. Already out of favor in the 1930s, by the 1950s and 1960s it was ancient history. The discoveries of Louis Leakey and his family all but proved that the first humans and their ancestors had been African and not Asian. Today the preponderance of evidence and scientific theory is firmly planted in one form or another of the 'out of Africa' scenario. Except for the concept of adaptive radiation (which is still considered a valid hypothesis), the particulars of his theory faded quickly, as did his reconciliationist view of science and religion. Geneticists and philosophers had grown increasingly less interested in bringing the two sides together to learn about evolution. A growing number of Christians joined the fundamentalists in their distaste for science. Osborn was like an envoy abandoned by parties for whom the idea of reconciliation to each other's points of view had grown irrelevant. Today, mainstream biologists, geologists and paleontologists no longer search for proofs of God or salvation, and aside from creationists, theologians are no longer interested in supporting their beliefs with physical evidence. Some writers such as Michael Behe, Philip Johnson and Michael Denton blend science and religion, or at least a thinly veiled spirituality, in the study of evolution and human origins. Though they argue for complexity from design and the guidance of an overreaching power and search for transcendent truths, they combine science and metaphysics in an attempt to show that naturalistic evolution is at best an intellectual construction and not the blind function of biology. Unlike Osborn's, the work of these creationists is anti-Darwinian. There seem to be no more romantic empiricists in the pro-evolution Osborn mold.

Though no anthropologists follow Osborn's human-origin theory at present, it does echo faintly in places. Michael Cremo and Richard Thompson's controversial non-ape ancestry theory argues that modern humans are as much as fifty-five million years old. According to their theory, humans appeared well before the advent of the primates and are separate species. As followers of Krishna Consciousness, Cremo and Thompson claim that the 'hidden history' of human evolution is written in Indian Vedic literature and mythology.[5]

Osborn can also be heard in the 'multiregional' hypotheses of another human-origins research team, Wolpoff and Caspari. Although they make an effort to separate their work from overtly racist polygenism, they argue that a population of proto-humans appeared in Africa about two million years ago and then quickly radiated into several branches which eventually became separate modern human species.[6] Osborn's multiple species idea was not completely discarded. A growing consensus amongst mainstream paleoanthropology is that various early people were different from one another. Though none argue for a special race plasm, it is generally believed that the Neanderthals, Cro-Magnons and Java Men were separate species, and not in a direct line from archaic to modern.

The physical legacy Osborn left behind is the museum itself. Until a multimillion-dollar renovation in the 1990s, the museum's fossil halls were a tribute to Osborn and Roy Chapman Andrews. Every dinosaur on display was either collected by Andrews or by an expedition sponsored by Osborn. Now far more politically correct (the Hall of the Age of Man is long gone), the museum is no longer a bastion of white male superiority. Some ghosts, however, may always linger.

The giant Theodore Roosevelt Memorial continues to tower over Central Park West and the millions of visitors who throng to the museum every year. Teddy sits astride his horse, native bearers faithfully at his side, grinning down on the world like another fossil from a bygone age. I like to think how proud Osborn would have been to walk past the statue every day, and how Roy Chapman Andrews would have heaved a sigh of regret that those glory days were over. William K. Gregory may have sighed as well, but at the pompous absurdity of it, while George Gaylord Simpson would have chuckled at the irony and got the joke that Osborn missed about lily-white adventurers who thought they were saving a multi-ethnic society.

Did Osborn prove his soul would be saved? Certain of so many things, even he was not sure. Henry Fairfield Osborn's search for the origins of man suggests to what extent all history and science is personal. Maybe all intellectual pursuit is the pursuit of ourselves and how we fit into the great complexity of the world regardless of what facts or evidence suggest. Osborn's work shows that questions of race, ethnicity and human relations are never simple. They are constructed from science, religion, culture and politics, and take into account not only how we treat each other, but how we treated each other in the distant past. Osborn claimed that knowledge of our origins would guide us in finding answers to those larger questions. On that he was correct.

Notes

1. *New York Times*, Obituary, 1935.
2. It can be argued that Osborn became the model for the doddering, but brilliant old professor character that worked its way into the science-fiction cinema of the 1950s, including such films as *Them* and *The Beast From 20 Thousand Fathoms*.
3. Charles R. Knight typescript letter to *American Magazine of Art*, 10/9/1935, NYPL.

4. Charles R. Knight to *New York Times*, n.d. (poss. 1947), NYPL.
5. Michael Cremo and Richard Thompson, *The Hidden History of the Human Race* (Badger, CA: Govardhan Hill Publishing, 1994).
6. Milford H. Wolpoff and Rachel Caspari, *Race and Human Evolution* (New York: Simon and Schuster, 1997).

Bibliography

Books (Primary)

Andrews, Roy Chapman, *Ends of the Earth*, New York: G.P. Putnam's Sons, 1929.

———, *The New Conquest of Central Asia*, New York: American Museum of Natural History, 1924.

———, *On the Trail of Ancient Man*, New York: G.P. Putnam's Sons, 1926.

———, *Under a Lucky Star*, New York: Viking Press, 1943.

Annual Report of the Board of Regents of the Smithsonian Institution, Washington: Government Printing Office, July 1890.

Barker, A.T., ed., *The Letters of H.P. Blavatsky to A.P. Sinnett*, London: T. Fisher Unwin, 1925.

Bartlett, John Russel, *The Progress of Ethnology, Archaeological, Philological and Geographical Researches in Various Parts of the Globe*, New York: Bartlett & Welford, 1847.

Bates, David Homer, *Lincoln in the Telegraph Office: Recollections of the United States Military Telegraph Corps during the Civil War* (New York: Century Co., 1907).

Bergson, Henri, *Creative Evolution*, New York: H. Holt & Co., 1907.

Blavatsky, H.P., *Isis Unveiled*, New York: Theosophical Publishing Society, 1877.

———, *The Secret Doctrine*, London: Theosophical Publishing Society, 1893.

Blumenbach, Johann Friedreich, *The Anthropological Treatises of Johann Friedreich Blumenbach*, London: Longman, Green, Longman, Roberts & Green, 1865.

Brigham, Carl C., *A Study of American Intelligence*, Princeton: Princeton University Press, 1923.

Bright, E., *Old Letters and Memories of Annie Besant*, Adyar: Theosophical Publishing House, 1936.

Cable, Mildred and Francesca French, *The Challenge of Central Asia,* New York: World Dominion Press, 1929.

Cope, E.D., *Origin of the Fittest: Essays in Evolution*, New York: Macmillan, 1877.

———, *Theology of Evolution, a Lecture*, Philadelphia: Arnold, 1887.

Darwin, Francis, ed., *The Life and Letters of Charles Darwin*, New York: Appleton's, 1903.

David-Neel, Alexandra, *My Journey to Lhasa*, New York: Harper & Bros, 1927.

Dixon, Ronald B., *The Racial History of Man*, New York: Scribner's, 1923.
Elliot-Smith, Grafton, *The Search for Man's Ancestors*, London: Watts and Co., 1932.
Foord, John, *The Life and Public Service of Andrew Haswell Green*, Garden City, NY: Doubleday Publishing, 1913.
Gobineau, Arthur de, *Inequality of the Races*,
———, *The Moral and Intellectual Diversity of the Races*, Philadelphia: J.B. Lippincott & Co., 1856.
Grant, Madison, *The Alien in Our Midst*, New York: Galton Pub. Co., 1930.
———, *The Conquest of the Continent*, New York: Scribner's Sons, 1933.
———, *Die Eroberung eines Kontinents*, German edn, Berlin: A. Metzner, 1937.
———, *Hank, His Lies and His Yarns*, New York: privately printed, 1931.
———, *The Passing of the Great Race*, New York: Scribner's Sons, 1916.
———, *The Vanished Game of Yesterday*, New York: 1933.
Grant, Madison and C.S. Davison, eds, *The Founders of the Republic on Immigration Naturalization and Aliens*, New York: Scribner's, 1928.
Gregory, William King, 'Biographical Memoir of Henry Fairfield Osborn, 1857–1935,' in *National Academy of Sciences Biographical Memoirs XIX*, 53–119.
Guyot, Arnold, *Earth and Man*, Boston: Gould and Lincoln, 1870.
Hooton, Earnest, *Twilight of Man*, New York: G.P. Putnam's Sons, 1939.
Hornaday, William Temple, *Our Vanishing Wildlife: Its Extermination and Preservation*, New York: New York Zoological Society, 1913.
Hrdlicka, Ales, *Early Man in South America*, Bureau of American Ethnology Bulletin 52, Washington: Government Printing Office, 1912.
———, *Skeletal Remains Suggesting or Attributed to Early Man in North America*, Bureau of American Ethnology Bulletin 33, Washington: Government Printing Office, 1907.
Kuhn, Alvin Boyd, *Theosophy: A Modern Revival of Ancient Wisdom*, New York: Holt & Co., 1930.
Leadbeater, C.W., *A Textbook of Theosophy* (13th edn), Madras, India: Theosophical Publishing House, 1979.
Morgan, T.H., *A Critique of the Theory of Evolution*, Princeton: Princeton University Press, 1916.
Osborn, Henry Fairfield, *The Age of Mammals*, New York: Macmillan, 1910.
———, *The American Museum of Natural History, its Origins and History*, New York: Irving Press, 1911.
———, *Cope: Master Naturalist*, New York: Princeton University Press, 1931.
———, *Creative Education in School, College, University, and Museum*, New York: Charles Scribner, 1927.

———, *The Earth Speaks to Bryan*, New York: Charles Scribner's Sons, 1925.

———, *Evolution and Religion*, New York: Charles Scribner's Sons, 1923.

———, *Evolution and Religion in Education: polemics in the fundamentalist controversy of 1922 to 1926*, New York: Scribner's, 1926.

———, *Fifty Two Years of Research Observation and Publication 1877–1929*, New York: Charles Scribner's Sons, 1930.

———, *The Hall of the Age of Man*, American Museum of Natural History Guidebook 52, New York: 1925.

———, *Hunting the Ancestral Elephant in the Fayum Desert*, New York: Century Co., 1907.

———, *Huxley and Education*, New York: Charles Scribner's Sons, 1910.

———, *Impressions of the Great Naturalists*, New York: Charles Scribner's Sons, 1924.

———, *Man Rises To Parnassus*, Princeton: Princeton University Press, 1927.

———, *Men of the Old Stone Age*, New York: Charles Scribner's Sons, 1925.

———, *The New Order of Sainthood*, New York: Scribner's, 1913.

———, *The Origin and Evolution of Life*, New York: Scribner's Sons, 1917.

———, *Science and Religion: Polemics in the Fundamentalist Controversy*, New York: Charles Scribner's Sons, 1926.

———, *The Titanotheres of Nebraska and Montana*, 1929.

Ripley, William Zebina, *The Races of Europe: a sociological study*, New York: Appleton & Co., 1889.

Roerich, Nicholas, *The Heart of Asia*, New York: Roerich Museum Press, 1930.

Scott, William Berryman, *Some Memories of a Paleontologist*, Princeton: Princeton University Press, 1939.

Smith, Samuel Stanhope, *An Essay on the Cause of the Variety of Complexion and Figure in the Human Species*, 2nd edn, New Brunswick: J. Simpson and Co., 1810.

Stoddard, Lothrop, *The Revolt against Civilization; the revolt of the under man*, New York: Scribner's Sons, 1922.

Straton, John Roach and Charles Francis Potter, *Evolution versus Creation*, New York: George H. Doran, 1924.

Whitman, Walt, *Leaves of Grass*, New York: 1855

Books (Secondary)

Abrams, M.H., *Natural Supernaturalism: Tradition and Revolution in Romantic Literature*, New York: W.W. Norton & Co., 1971.

Ahlstrom, Sydney, *A Religious History of the American People*, New Haven: Yale University Press, 1974.

Allen, Garland, *Life Science in the Twentieth Century*, Cambridge: Cambridge University Press, 1975.

Allen, Oliver, *New York, New York*, New York: Macmillan Publishing Co., 1990.

Allen, Don Cameron, *The Legend of Noah*, Urbana: University of Illinois Press, 1949.

Barkan, Elazar, *The Retreat of Scientific Racism*, New York: Cambridge University Press, 1992.

Benson, Keith R., Jane Maienschein and Ronald Rainger, eds, *The Expansion of American Science*, New Brunswick: Rutgers University Press, 1991.

Berg, A. Scott, *Max Perkins, Editor of Genius*, New York: Washington Square Press, 1979.

Bowler, Peter J., *Eclipse of Darwinism: Anti-Darwinian Evolutionary Theories in the Decades around 1900*, Baltimore: Johns Hopkins University Press, 1983.

———, *Evolution: The History of an Idea*, Berkeley: University of California Press, 1983.

———, *Life's Splendid Drama*, Chicago: University of Chicago Press, 1996.

———, *Theories of Human Evolution: a century of debates, 1844–1944*, Baltimore: Johns Hopkins University Press, 1986.

Blinderman, Charles, *The Piltdown Inquest*, Buffalo, New York: Prometheus Books, 1986.

Bloor, David, *Knowledge and Social Imagery*, Chicago: University of Chicago Press, 1991.

Bridges, William, *Gathering of Animals: An Unconventional History of the New York Zoological Society*, New York: Harper & Row, 1974.

Broderick, Alan Houghton, *The Abbé Breuil, Prehistorian*, London: Hutchinson, 1963.

Brooke, John Hedley, *Science and Religion*. Cambridge: Cambridge University Press, 1993.

Carey, John, *Intellectuals and the Masses*, New York: 1992.

Chase, Allen, *The Legacy of Malthus*, New York: Alfred A. Knopf, 1977.

Conkin, Paul, *When All the Gods Trembled: Darwinism, Scopes, and American Intellectuals*, New York: Rowman and Littlefield Publishing Co., 1998.

Cowley, Malcolm, *The Dream of the Golden Mountains*, New York: Penguin Books, 1964.

Cravens, Hamilton, *The Triumph of Evolution: American Scientists and the*

Heredity–Environment Controversy, 1900–1941, Philadelphia: University of Pennsylvania Press, 1978.

Cremo, Michael A. and Richard L. Thompson, *The Hidden History of the Human Race*, Badger, CA: Govardhan Hill Publishing, 1994.

Cudahy, Brian J., *Under the Sidewalks of New York: The Story of the Greatest Subway System in the World*, New York: Fordham University Press, 1991.

Czerkas, Sylvia and Donald F. Glut, *Dinosaurs, Mammoths, and Cavemen: the Art of Charles R. Knight*, United States: E.P. Dutton, Inc., 1982.

Daniels, George H., *American Science in the Age of Jackson*, Tuscaloosa: University of Alabama Press, 1968.

De Camp, S.L., *Lost Continents: The Atlantis Theme in History, Science, and Literature*, New York: Gnome Press, 1954.

Deichmann, Ute, *Biologists Under Hitler*, Cambridge, MA: Harvard University Press, 1996.

Degler, Carl, *In Search of Human Nature*, New York: Oxford University Press, 1991.

Douglas, Ann, *Terrible Honesty: Mongrel Manhattan in the 1920s*, New York: Noonday Press, 1995.

Dupree, A. Hunter, *Asa Gray: American Botanist, Friend of Darwin*, Baltimore: Johns Hopkins University Press, 1988.

Fuller, Robert, *Henry Fairfield Osborn: The Man and His Books*, New York: Scribner's Co., n.d.

Gallenkamp, Charles, *Dragon Hunter: Roy Chapman Andrews and the Central Asiatic Expeditions*, New York: Viking, 2001.

Gasman, Daniel, *The Scientific Origins of National Socialism*, New York and London: Macdonald & Co., 1971.

———, *Haeckel's Monism and the Birth of Fascist Ideology*, Peter Land Publishing, 1998.

Gilber, James Burkhart, *Redeeming Culture: American Religion in an Age of Science*, Chicago: University of Chicago Press, 1997.

Glick, Thomas F. and David Kohn, eds, *Charles Darwin On Evolution*, Indianapolis: Hackett Publishing Co., 1996.

Godwin, Joscelyn, *Arktos: The Polar Myth in Science, Symbolism, and Nazi Survival*, Kempton, IL: Adventures Unlimited Press, 1996.

Goodrick-Clark, Nicholas, *The Occult Roots of Nazism*, New York: New York University Press, 1992.

Gould, Stephen Jay, *Ontogeny and Phylogeny*, Cambridge, MA: Belknap Press, 1977.

———, *The Mismeasure of Man*, New York: W.W. Norton & Co., 1996.

Guiley, Rosemary Ellen, *Harper's Encyclopedia of Mystical and Paranormal Experience*, New York: Harper, 1991.

Haller, John S. Jr, *Outcasts from Evolution: Scientific Attitudes to Racial Inferiority, 1859–1900*, Urbana: University of Illinois Press, 1971.

Haller, Mark, *Eugenics: Hereditary Attitudes in American Thought*, New Brunswick: Rutgers University Press, 1963.

Haraway, Donna, *Primate Visions*, New York: Routledge, 1989.

Hare, Harold Edward and William Loftus Hare, *Who Wrote the Mahatma Letters?*, London: Williams and Norgate Ltd, 1936.

Harwood, Jonathan, *Styles of Scientific Thought: the German Genetics Community 1900–1933*, Chicago: University of Chicago Press, 1993.

Hayes, Samuel P., *The Response to Industrialism: 1885–1914*, Chicago: University of Chicago Press, 1957.

Higham, John, *Strangers in the Land: Patterns of Nativism 1860–1925*, New York: Atheneum, 1973.

Hindle, Brooke, *The Pursuit of Science in Revolutionary America 1735–1789*, Chapel Hill: University of North Carolina Press, 1956.

Hofstader, Richard, *The Age of Reform*, New York: Vintage Press, 1955.

———, *Social Darwinism in American Thought*, Boston: Beacon Press, 1955.

Howe, Elic, *Magicians of the Golden Dawn*, Wellingborough: Aquarian Press, 1985.

Howe, Irving, *World of Our Fathers*, New York: Harcourt, Brace, Jovanovich, 1976.

Hughes, Robert, *American Visions: The Epic History of Art in America*, New York: Alfred Knopf, 1997.

Hutchinson, William, *The Modernist Impulse in American Protestantism*, Cambridge: Harvard University Press, 1976.

Kerr, Howard and Charles L. Crow, eds, *The Occult in America: New Historical Perspectives*, Urbana: University of Illinois Press, 1986.

Kevles, Daniel J., *In the Name of Eugenics*, New York: Alfred Knopf, 1985.

Koenig, Louis W., *Bryan: A Political Biography of William Jennings Bryan*, New York: Putnam, 1971.

Kohl and Fawcett, *Nationalism, Politics and the Practice of Archeology*, New York: Cambridge University Press, 1995.

Kuhl, Stefan, *The Nazi Connection*, New York: Oxford University Press, 1994.

Landau, Misia, *Narratives of Human Evolution*, New Haven: Yale University Press, 1991.

Lanham, Url, *The Bone Hunters*. New York: Columbia University Press, 1973.

Lanpo, Jia, *Early Man in China*, Beijing: Foreign Language Press, 1980.

Larson, Edward, *Sex, Race, and Science*,

———, *Summer For the Gods*, Cambridge: Harvard University Press, 1997.

Lewin, Roger, *Bones of Contention*, New York: Touchstone, 1987.

Lewis, R.W.B., *American Adam*, Chicago: University of Chicago Press, 1955.

Ludmerer, Kenneth M., *Genetics and American Society: a historical appraisal*, Baltimore: Johns Hopkins Press, 1972.

Marsden, George, *Fundamentalism and American Culture: The Shaping of the Twentieth-Century Evangelicalism 1870–1925*, New York: Oxford University Press, 1980.

Meyer, Karl and Shareen Blair Brysac, *Tournament of Shadows: The Great Game and the Race for Empire in Central Asia*, Washington, DC: Counterpoint, 1999.

Miller, Robert M., *Harry Emerson Fosdick: Preacher, Pastor, Prophet*, New York: Oxford University Press, 1985.

Mohr, James C., *The Radical Republicans and Reform in New York During Reconstruction*, Ithaca: Cornell University Press, 1973.

Moore, James, *Post-Darwinian Controversies*, New York: Cambridge University Press, 1979.

Morris, Edmund, *The Rise of Theodore Roosevelt*, New York: Ballantine Books, 1979.

Numbers, Ronald, *The Creationists*, Berkeley: University of California Press, 1992.

———, *Darwinism Comes to America*, Cambridge: Harvard University Press, 1998.

Paul, Diane B., *Controlling Human Heredity: 1865 to the Present,* New Jersey: Humanities Press, 1995.

———, *The Politics of Heredity*, Albany: State University of New York Press, 1998.

Pauwels, Louis and Jacques Bergier, *Morning of the Magicians*, New York: Stein & Day Publishers, 1964.

Poliakov, Leon, *The Aryan Myth*, New York: Basic Books, 1974.

Preston, Douglas J., *Dinosaurs in the Attic*, New York: St Martin's Press, 1986.

Proctor, Robert N., *Racial Hygiene: medicine under the Nazis*, Cambridge: Harvard University Press, 1988.

Rafferty, Edward, *Apostle of Human Progress: The Life of Lester Frank Ward*, Ann Arbor, MI: UMI, doc. diss., 1999.

Rainger, Ronald, *An Agenda for Antiquity: Henry Fairfield Osborn and Vertebrate Paleontology at the American Museum of Natural History, 1890–1935*, Tuscaloosa: University of Alabama Press, 1991.

Rainger, Ronald, Keith Benson and Jane Maienschien, *The American Development of Biology*, New Brunswick: Rutgers University Press, 1991.

Reiger, John, *Passing of the Great West: Selected Papers of George Bird Grinell*, New York: Scribner's, 1972.

Reingold, Nathan, *Science American Style*, New Brunswick: Rutgers University Press, 1991.

Ripley, Ernestine, *Bibliography of Published Writings of Henry Fairfield Osborn*, Lancaster, PA: private printing, 1916.

Roberts, Jon, *Darwinism and the Divine: Protestant Intellectauals and Organic Evolution*, Madison, WI: University of Wisconsin Press, 1988.

Rose, Anne, *American Victorians and the Civil War*, Cambridge: Cambridge University Press, 1992.

Rudwick, Martin, *Scenes From Deep Time*, Chicago: University of Chicago Press, 1992.

Ruse, Michael, *Monad to Man: the concept of progress in evolutionary biology*, Cambridge, MA: Harvard University Press, 1997.

Russett, Cynthia Eagle, *Darwin in America: The Intellectual Response 1865–1912*, San Francisco: W.H. Freeman & Co., 1976.

Said, Edward, *Orientalism*, New York: Vintage Books, 1979.

Sandeen, Ernest, *The Roots of Fundamentalism: British and American Millenarianism, 1800–1930*, Chicago: 1970.

Shipman, Pat, *The Evolution of Racism*, New York: Simon & Schuster, 1994.

———, *The Man Who Found the Missing Link*, New York: Simon Schuster, 2001.

Shipman, Pat and Erik Trinkhaus, *The Neanderthals: Changing the Image of Mankind*, New York: Alfred A. Knopf, 1993.

Shreeve, James, *The Neanderthal Enigma: Solving the Mystery of Modern Human Origins*, New York: William Morrow & Co. Inc., 1995.

Simpson, George Gaylord, *Concession to the Improbable: An Unconventional Autobiography*, New Haven: Yale University Press, 1978.

Slaughter, Thomas P., *The Natures of John and William Bartram*, New York: Alfred A. Knopf, 1996.

Sloan, G.R., *Geopolitics in United States Policy, 1890–1987*, New York: St Martin's Press, 1988.

Snyder, Louis L., *Race: A History of Modern Ethnic Theories*, New York: 1939.

Solomon, Barbara, *Ancestors and Immigrants*, Cambridge, MA: Harvard University Press, 1956.

Spiro, Jonathan, *Patrician Racist: The Evolution of Madison Grant*, University of California at Berkeley: Ph.D. diss., 2000.

Stepan, Nancy, *The Idea of Race in Science: Great Britain 1800–1960*, Hamden, CT: Archon Books, 1982.

Stocking, George W., ed., *Bones, Bodies, Behavior: essays on biological anthropology*, Madison: University of Wisconsin Press, 1988.

———, *Race, Culture and Evolution*, Chicago: University of Chicago Press, 1982.

Turner, Frank, *Between Science and Religion*, New Haven: Yale University Press, 1974.

Van Riper, A. Bowdoin, *Men Among the Mammoths: Victorian science and the discovery of human prehistory*, Chicago: University of Chicago Press, 1993.

Warren, Leonard, *Joseph Leidy: The Last Man Who Knew Everything*, New Haven: Yale University Press, 1998.

Washington, Peter, *Madame Blavatsky's Baboon*, New York: Schocken Books, 1995.

West, Robert Howard, *The Dawnseekers*, New York: Harcourt, Brace, Jovanovich, 1975.

Wiebe, Robert H., *The Search For Order*, New York: Hill and Wang, 1967.

Williams, Gertrude Leavenworth Marvin, *Priestess of the Occult, Madame Blavatsky*, New York: A.A. Knopf, 1946.

Wimmel, Kenneth, *The Alluring Target: in Search of the Secrets of Central Asia*, Washington: Trackless Sands Press, 1996.

Wolpoff, M.H. and Caspari, R., *Race and Human Evolution*, New York: Simon and Schuster, 1997.

Wyllie, Irvin, *The Self-Made Man in America: the Myth of Rags to Riches*, New Brunswick: Rutgers University Press, 1954.

Articles (Primary)

Acton, Harry, 'On the Gangplank,' *New York American* (October 9, 1931).

Ameghino, Florentino, 'L'age des formations sedimentaires de Patagonia,' *Anales de la Sociadad Cientific Argentina*, 50 (1900), 109–130, 145–165, 209–229; 51 (1901).

Baldwin, J.M. 'A New Factor in Evolution,' *American Naturalist*, 30:354 (June 1896), 441–533.

Barbour, Erwin Hinckley and Henry Baldwin Ward, 'Preliminary Report on the Primitive Man of Nebraska,' *Nebraska Geological Survey* (December 22, 1906).

Berkey, Charles and Nels Nelson, 'Geology and Prehistoric Archaeology of the Gobi Desert,' *Novitates* (March 18, 1926).

Cope, E.D., 'The Marriage Problem,' *The Open Court*, 64 (November 15, 1888), 1307–1324.

———, 'Two Perils of the Indo-European,' *The Open Court*, 3 (1890), 2052–2054.

———, 'The Relation of the Sexes to Government,' *Popular Science Monthly* (October 1888), 730.

Dart, Raymond, 'Australopithecus africanus: The Man-Ape of South Africa,' *Nature*, 2884:115 (February 7, 1925), 195–199.

Frazer, Persifor, 'The Life and Letters of Edward Drinker Cope,' *American Geologist*, 26:2 (August 1900), 67–128.

Grant, Madison, 'America for the Americans,' *The Forum*, 74:3 (1925), 346-355.

———, 'History of the Zoological Society,' *Zoological Society Bulletin*, 37 (January 1910).

———, 'The Origin of the New York Zoological Society,' in George Bird Grinell and Theodore Roosevelt, eds, *Trail and Campfire: The Book of the Boone and Crockett Club*, New York: 1897.

———, 'The Racial Transformation of America,' *North American Review*, 219 (1924), 345–352.

Gregory, William King, 'A Critique of Professor Osborn's Theory of Evolution,' *American Journal of Physical Anthropology*, 13 (1930), 133–164.

———, 'Biographical Memoir of Henry Fairfield Osborn. 1857–1935,' *National Academy of Sciences Biographical Memoirs XIX*, Washington: National Academy of Sciences (1938), 53–119.

———, 'The Dawn Man of Piltdown, England,' *American Museum Journal*, 14 (1914), 189–200.

———, 'Is the Pro-Dawn Man a Myth?', *Human Biology*, 1 (May 1929), 153–165.

———, 'Studies on the Evolution of Primates: Phylogeny of Recent and Extinct Anthropoids with Special Reference to the Origin of Man,' *Bulletin of the American Museum of Natural History*, 35 (1916), 258–355.

———, 'Two Views in the Origin of Man,' *Science*, lxv:1695 (June 24, 1927), 601–605.

Gregory, William K. and Milo Hellman, 'Notes on the Type of Hesperopithecus Harold Cookii Osborn,' *American Museum Novitates*, 53 (January 6, 1923).

Keith, Arthur, Grafton Elliot-Smith and Arthur Smith-Woodward, 'The Fossil Anthropoid Ape of Taung,' *Nature*, 2885:115 (February 14, 1925), 234–236.

Lull, Richard Swann, 'The Antiquity of Man,' in *The Evolution of Man*, New Haven: Yale University Press, 1921.

Matthew, W.D., 'Hypothetical Outlines of the Continents in Tertiary Times,' *Bulletin of the American Museum of Natural History*, 22 (1906), 353–383.

———, 'A Provisional Classification of the Freshwater Tertiary of the West,' *Bulletin of the American Museum of Natural History*, 12 (1899), 19–75.

———, 'New Canidae from the Miocene of Colorado,' *Bulletin of the American Museum of Natural History*, 16 (1902), 286-287.

Matthew, W.D. and Harold Cook, 'Pliocene Fauna from Western Nebraska,' *Bulletin of the American Museum of Natural History*, 26 (1909), 361–414.

Mencken, H.L., 'Theodore Roosevelt, Pseudo Liberal,' in *A Mencken Chrestomathy*, New York: Alfred Knopf, 1920.

Morgan, C.L., 'On Modification and Variation,' *Science*, n.s., 4:99 (November 20, 1896), 733–740.

Osborn, Henry Fairfield, 'Address of Welcome,' Second International Congress of Eugenics, Eugenics and the Family I, Baltimore: Williams & Wilkins Company, 1923.

———, 'The American Museum And Citizenship,' 54th Annual Report of the Trustees of the American Museum of Natural History for the year 1922 (1923).

———, 'Are Acquired Variations Inherited?,' *American Naturalist*, 25 (March 1891), 291–216.

———, 'Aristogenesis, the Creative Principle in the Origin of Species,' *American Naturalist*, 68 (1934), 193–234.

———, 'Biological Inductions From the Evolution of the Probocidea,' *Science*, 76:1979 (December 2, 1932), 501–504.

———, 'The Cavemen Knew,' *Collier's Magazine*, 75:21 (May 23, 1925), 23.

———, 'Coincident Evolution Through Rectigradations and Fluctuations,' *Science*, 27 (1908), 749–752.

———, 'The Continuous Origin of Certain Unit Characters as Observed by a Paleontologist,' *American Naturalist*, 46 (April 1912), 185–206.

———, 'Discovery of a Supposed Primitive Race of Men in Nebraska,' *Century Magazine* (January 1907), 371–375.

——— 'Evolution as it Appears to the Paleontologist,' *Science*, 26 (1907), 744–749.

———, 'The Evolution of the Horse in America,' *Century Magazine*, LXIX (November 1904), 3–17.

———, 'Evolution of Mammalian Molars to and from the Tritubercular Type,' *American Naturalist*, 22:264 (December 1888), 1067–1079.

———, 'The Fighting Ability of the Different Races,' *Journal of Heredity*, 10:1 (1919), 29–31.

———, 'The Four Inseparable Factors of Evolution,' *Science*, n.s., 27:682 (January 24, 1908), 148–150.

———, 'The Geological and Faunal Relations of Europe and America During the Tertiary Period and the Theory of the Successive Invasions of an African Fauna,' *Science*, n.s., 11:276 (April 13, 1900), 561–574.

———, 'The Hereditary Mechanism and the Search for the Unknown Factors of Evolution,' *American Naturalist* (May 1895), 418–429.

———, 'Hesperopithecus, The Anthropoid Primate of Western Nebraska,' *Nature*, 110:2750 (August 26, 1922), 281–283.

———, 'Hesperopithecus, The First Anthropoid Primate Found in America,' *American Museum Novitates*, 37 (April 25, 1922).

———, 'How to Teach Evolution in Schools,' *School and Society*, 23 (1926), 25–31.

———, 'Is the Ape-man a Myth?,' *Human Biology*, 1 (January 1929), 4–9.

———, 'The Law of Adaptive Radiation,' *American Naturalist*, 36:425 (May 1902), 353–363.

———, 'Lo, The Poor Nordic,' *New York Times* (4/8/1924).

———, 'Mutations of Waagen and "Mutations" of De Vries or "Rectigradations" of Osborn,' *Science*, 33 (March 3, 1911), 328.

———, 'Nomenclature of Mammalian Molar Cusps,' *American Naturalist*, 22:262 (October 1888), 926–928.

———, 'Ontogenetic and Phylogenetic Variation,' *Science*, n.s., 4:100 (November 1897), 789.

———, 'Opening Address,' in *Eugenics, Genetics, and the Family, Volume I, Second International Congress on Eugenics*, Baltimore: Williams & Wilkins Co., 1923, preface.

———, 'Opening of the Hall of Public Health,' in *Articles on J.P. Morgan and the American Museum*, New York: *American Museum Journal*, 1913.

———, 'Origin of the Mammalia,' *American Naturalist*, 32:377 (May 1898), 309–334.

———, 'Organic Selection,' *Science*, n.s., 6:146, 585 (October 1897), 583–587.

———, 'Orthogenesis as Observed From Paleontological Evidence Beginning in the Year 1889,' *American Naturalist*, 56 (1922), 134–143.

———, 'The Paleontological Evidence for the Transmission of Acquired Characters,' *Nature* (January 9, 1890), 227–228.

———, 'Paleontology Versus Devriesianism and Genetics in the Factors of the Evolution Problem,' *Science*, 73:1899 (May 22, 1931), 547–549.

———, 'Present Problems in Heredity,' *Atlantic Monthly*, 67 (March 1891), 353–364.

———, 'The Principles of Evolution Revealed by Paleontology,' *Nature*, 3239:128 (November 28, 1931), 894–897.

———, 'Recent Discoveries Relating to the Origin and Antiquity of Man,' *Science*, 65 (May 20, 1927), 481–488.

———, 'Search for the Unknown Factors in Evolution,' *American Naturalist* (May 1895), 418–439.

———, 'Selecting the American Child,' *Kindergarten and First Grade Magazine*, 10 (September 1924), 5–7.

———, 'The Seven Factors of Education,' *Educational Review*, 32 (1906), 56–82.

———, 'Why Central Asia?,' *Natural History*, 26:3 (1926), 263–269.

Parker, Dorothy, 'Arrangement in Black and White,' *The Portable Dorothy Parker*, New York: Penguin Books, 1976.

Peake, H.J.E, 'Review of *The Passing of The Great Race*,' *Man*, 20:173 (November 1920).

Pearl, Raymond, 'The Biology of Superiority,' *American Mercury*, 12 (November 1927), 257–266.

Ripley, William, Z., 'The Racial Geography of Europe,' *Popular Science Monthly*, L (March 1897), 577–594.

Roosevelt, Theodore, 'Race Decadence,' *Outlook*, 97 (1911), 763–769.

Scott, William Berryman, 'On Variations and Mutations,' *American Journal of Science*, 3rd edn, 48 (1894), 355–374.

Simpson, George Gaylord, 'Osborn Biography,' *Dictionary of American Biography*, New York: 1944.

Suess, ed., 1899, 'Uberreste Von Rhinoceros sp. Aus der Ostlichen Mongolei,' Mit Anmerkungen Von W. Obrutschen, Ver Handl. KK. Mineral. Gesell. *St Petersburg, Zweite Series*, Band XXXVI, no. 2.

Articles (Secondary)

Alexander, Charles C., 'Prophet of American Racism: Madison Grant and the Nordic Myth,' *Phylon*, 23:1 (1962), 73–90.

Allen, Garland E., 'The Eugenics Record Office at Cold Spring Harbor, 1910–1940, An Essay in Institutional History,' *Osiris*, 2nd series, 2 (1986), 225–264.

———, 'The Misuses of Biological Hierarchies: the American eugenics movement 1900–1940,' *History and Philosophy of Life Sciences*, 5 (1984), 105–128.

———, 'T.H. Morgan and the Emergence of a New American Biology,' *Quarterly Review of Biology*, 44 (1969), 168–188.

———, 'Thomas Hunt Morgan: Materialism and Experimentalism in the development of Modern Genetics,' *Social Research*, 51:3 (Autumn 1984), 707–738.

Allworth, Edward A., 'The Rediscovery of Central Asia: the Region Reflected in Two Collections of the New York Public Library,' *Biblion* (1995), 95–125.

Blanckaert, Claude, 'On the Origins of French Ethnology,' in George W. Stocking, ed., *Bones, Bodies, Behavior*, Madison: University of Wisconsin Press, 1988.

Boaz, Noel, 'American Research on Australopithecine and Early Homo, 1925–1980,' in Frank Spencer, ed., *A History of American Physical Anthropology*, New York: Academic Press, 1982, pp. 240–241.

Bowler, Peter, 'Edward Drinker Cope and the Changing Structure of Evolutionary Thought,' *Isis*, 68:242 (1977), 255.

Brechin, Gray, 'Conserving the Race: Natural Aristocracies, Eugenics, and the U.S. Conservation Movement.' *Antipode*, 28:3 (1996), 229–245.

Colbert, Edwin H., 'Biographical Memoir of William K. Gregory,' *National Academy of Science*, 46 (1975), 91–133.

———, 'Four Giants of Paleontology,' *Natural History*, 5 (1994), 62–67.

Dayton, Donald, 'Dispensationalism and the Emergence of Fundamentalism Among American Baptists,' Keynote Address, *Methodism and the Fragmentation of American Protestantism*, Northern Baptist Theological Seminary (September 29, 1995).

Fairchild, Halford H., 'Scientific Racism: The Cloak of Objectivity,' *Journal of Social Issues*, 47:3 (1991), 101–115.

Gidley, M., 'Notes on F. Scott Fitzgerald and the Passing of the Great Race,' *American Studies*, 7:2 (1973), 171–181.

Gould, Stephen Jay, 'An Essay on a Pig Roast,' *Natural History* (June 1989), 14–25.

Hammond, Michael, 'The Shadow Man Paradigm in Paleoanthropology 1911–1945,' in George W. Stocking, ed., *Bones, Bodies, Behavior: Essays on Biological Anthropology*, Madison: University of Wisconsin Press, 1988, pp. 117–137.

Hays, Samuel, 'The Politics of Reform in Municipal Government in the Progressive Era,' *Pacific Northwest Quarterly*, lv (October 1964), 157–169.

Higham, John, 'A Brake on Nativism,' in Arthur Mann, ed., *The Progressive Era*, New York: Rinehart & Winston, 1963.

Horowitz, Helen, 'Animal and Man in the New York Zoological Park,' *New York History*, 55 (1975), 426-453.

Kennedy, Michael,'Philanthropy and Science in New York City: the American Museum of Natural History, 1868–1968,' Ph.D. diss., Yale University, 1968.

Kottler, Malcolm Jay, 'Alfred Russel Wallace, the Origin of Man, and Spiritualism,' *Isis*, 65 (1974), 145–192.

Moore, James, 'Herbert Spencer's Henchmen: The Evolution of Protestant Liberals in late Nineteenth-century America,' in John Durant, ed., *Darwinism and Divinity*, NewYork: Basil Blackwell, 1985, pp. 76-99.

Mosedale, Susan Sleeth, 'Science Corrupted: Victorian Biologists Consider "The Woman Question,"' *Journal of the History of Biology*, 11:1 (Spring 1978), 1–55.

Paul, Diane B. and Barbara A Kimmelman, 'Mendel in America: Theory and Practice, 1900–1919,' in Ronald Rainger, Ronald Benson and Jane Maienschein, eds, *The American Development of Biology*, Philadelphia: University of Philadelphia Press, 1988, pp. 281–310.

Pfeifer, Edward J., 'Genesis of American Neo-Lamarckism,' *Isis*, 56 (1965), 156-167.

Potter, Charlotte, 'The Rise to Parnassus: Henry Fairfield Osborn and the Hall of the Age of Man,' *Museum Studies Journal* (Spring 1983), 26-34.

Preston, Douglas, 'A Daring Gamble in the Gobi Desert Took the Jackpot,' *Smithsonian Magazine* (December 1987), 94–105.

Radford, John, 'Sterilization Versus Segregation: Control of the "Feebleminded" 1900–1938,' *Society of Science and Medicine*, 33:4 (1991), 449–458.

Rainger, Ronald, 'Just Before Simpson: William Diller Matthew's Understanding of Evolution,' *Proceedings of the American Philosophical Society*, 130 (1986).

———, 'Vertebrate Paleontology as Biology: Henry Fairfield Osborn and the American Museum of Natural History,' in Ronald Rainger, Keith R. Benson and Jane Maienschein, eds, *The American Development of Biology*, Philadelphia: University of Philadelphia Press, 1988, pp. 219–265.

———, 'What's The Use: William King Gregory and the Functional Morphology of Fossil Vertebrates,' *Journal of the History of Biology*, 22:1 (Spring 1989), 103–139.

Shapiro, Harry, 'In Memory of Frederick H. Osborn,' *Social Biology*, 29 (1982), i–iii.

Sloan, Douglas, 'Science in New York City 1867–1907,' *Isis*, 71 (1980), 35–76.

Swierenga, Robert P., 'Religion and Immigration Behavior: The Dutch Experience,' in Philip R. Brook and Robert P. Swierenga, eds, *Belief and Behavior: Essays in the New Religious History*, New Brunswick: Rutgers University Press, 1991.

Turlish, Lewis A., '*The Rising Tide of Color*: A Note on the Historicism of *The Great Gatsby*,' *American Literature*, 43 (Autumn 1971), 442–444.

Wolf, John and James S. Mellett, 'The Role of "Nebraska Man" in the Creation/Evolution Debate,' *Creation/Evolution*, 16 (1985), 31–43.

Index